Making Fishery Agreements Work

Making Fishery Agreements Work

Post-Agreement Bargaining in the Barents Sea

Geir Hønneland

Fridtjof Nansen Institute and the University of Tromsø, Norway

Edward Elgar

Cheltenham, UK • Northampton, MA, USA

Published by
Edward Elgar Publishing Limited
The Lypiatts
15 Lansdown Road
Cheltenham
Glos GL50 2JA
UK

Edward Elgar Publishing, Inc.
William Pratt House
9 Dewey Court
Northampton
Massachusetts 01060
USA

A catalogue record for this book
is available from the British Library

Library of Congress Control Number: 2011939351

MIX
Paper from
responsible sources
FSC
www.fsc.org   FSC® C018575

ISBN 978 0 85793 362 1 (cased)

Typeset by Servis Filmsetting Ltd, Stockport, Cheshire
Printed and bound by MPG Books Group, UK

Contents

Acknowledgements

This book was written with financial support from the programme 'The Oceans and Coastal Areas (HAVKYST)' under the Research Council of Norway. Thanks to the project's Ph.D. student Anne-Kristin Jørgensen for numerous discussions about the topics raised in the book, to Stig Gezelius and Olav Schram Stokke for thorough and constructive comments, to Susan Høivik for splendid language-editing, to Claes Lykke Ragner and Maryanne Rygg for excellent technical assistance, and to Kari Lorentzen for efficient library service. The first half of the book was written during a visit to Innovative Fisheries Management (IFM) at the University of Aalborg in summer 2010. Thanks to Jesper Raakjær for accepting me and Anne-Kristin as guest researchers and to the staff at IFM in Hirtshals for making our stay memorable. Finally, thanks to all the Norwegian and Russian civil servants and scientists who took the time to let me interview them.

1. Introduction

Why do people obey the law? And why do states abide by their international commitments? The concept of compliance has attracted increasing interest among political scientists in recent years. The debate relates to political processes at various levels involving many types of subjects. At the international level, the focus is on how states comply with international treaties and regime obligations; at the national level, on how lower-level bodies deal with decisions made at higher levels; and at the individual level, on how specific individuals comply with rules aimed at regulating their behaviour. This literature builds on explorations and findings from economics, psychology, criminology and other social sciences. It generally has two main aims: to explain *why* subjects comply with certain regulations, and *how* the relevant authorities can enhance compliance. Why do drivers sometimes follow traffic rules, and fishers (again, sometimes) keep their catches within quota limits? Is it a matter of personal ethics, of economic calculations or of the legitimacy of political institutions? What types of political action can best nurture a person's inclination to comply? Why do states sometimes abide by the agreements they conclude with other states, and sometimes not? Do states have a moral or ethical sense? Do they fear shaming or retaliation from other states if they fail to keep their commitments? What strategies can states apply to get other states to stick to their promises?

This book looks into these questions by focusing on the management of one international fishery, examining compliance at both the state and the individual levels. In the process, we touch on issues like East–West communication and coordination in the European High North, where Russia is enmeshed in a web of collaborative networks with its Nordic neighbours. The setting is the Barents Sea, home to some of the most productive fishing grounds on the planet, including the world's largest cod stock. Since the 200-mile exclusive economic zones (EEZs) were introduced in the mid-1970s, Norway and the Soviet Union/the Russian Federation have managed the major fish stocks in the area together, through the Joint Norwegian–Russian Fisheries Commission. Some three and a half decades later, this bilateral management regime would appear to be a successful exception to the rule of failed fisheries management: stocks are in good shape; moreover, institutional cooperation is expanding, and takes place in a generally friendly atmosphere. Both parties present their accomplishments in the Barents Sea as an example for emulation.

Compliance with the bilateral agreements and the national fisheries regulations has, on the whole, been acceptable. Nevertheless, unsatisfactory compliance by Russian fishers has been a main topic for the Joint Commission since the early 1990s. Norway has repeatedly claimed that the Russians have overfished their quotas, and has employed a range of negotiation strategies to induce Russian fishers to comply with the agreed quota levels, and to get the Russian authorities to take the problem seriously. Norway has also continuously pressed Russia to agree on regulatory measures that can increase the parties' compliance with the internationally recognized precautionary approach to fisheries management. What kinds of strategies has Norway employed in this pursuit of compliance, and how has the Russian side perceived them? Can we find empirical support for theories of compliance in the encounters between Norwegian inspectors and Russian fishers, and, more broadly, in fisheries relations between the Eastern great power Russia and the Western small state Norway? This is a study of compliance at both the individual level and the state level, with a focus on how post-agreement bargaining can enhance compliance.

THEORETICAL POINTS OF DEPARTURE

In the economic literature, compliance with the law has largely been viewed as the result of cost–benefit calculations on the part of individuals, and of deterrence on the part of the public authorities: basically, people comply because they see it as being in their best interest, and because they fear punishment if they are detected in criminal acts.

In his influential book *Why People Obey the Law* from 1990 (re-issued in 2006), Tom R. Tyler provides compelling evidence that this is not necessarily the case, building on a long tradition in sociology. He holds that personal morality and the legitimacy enjoyed by the public authorities account better for compliance than does self-interest. People comply with the law when they feel that the law is fair and just, or when they feel obliged to follow the regulations made by a political system that they trust. In practice, people obey laws even when the probability of punishment for non-compliance is close to zero; and they break laws even in cases involving substantial risk. People may comply with a law that reflects their personal convictions, but violate laws that make less sense to them, economic calculations aside. For the public authorities, Tyler argues (2006, pp. 25–6), legitimacy provides a far more stable base for compliance than does morality. Such legitimacy rests on the feeling of obligation to obey any commands that an authority issues. It is 'a reservoir of loyalty on which leaders can draw, giving them the discretionary authority they require to govern effectively' (ibid., p. 26). In his own 'Chicago Study', analysing the experiences, attitudes and behaviour of a random sample of Chicagoans as regards everyday crime, Tyler finds legitimacy more influential than the risk of being caught and punished for breaking the rules. As he argues, 'people's motivation to cooperate with others, in this case legal authorities, is rooted in social relationships and ethical judgments, and does not primarily flow from the desire to avoid punishments or gain rewards' (ibid., p. 270). Tyler contrasts the instrumental (deterrence-based) approach to compliance to what he terms the normative perspective.

The motivation to cooperate is at the heart of the literature on how to regulate common-pool resources – or 'the commons'. These are resources where the acquisition of resource-units by one user takes place at the expense of other potential users – typically grazing lands, fish stocks, groundwater basins, irrigation canals and other bodies of water. In his seminal article 'The Tragedy of the Commons', Garrett Hardin (1968) describes a situation in which actors behaving independently and in the pursuit of self-interest will eventually bring a shared resource to extinction, even though that is not in the long-term interest of any of the actors. To avoid this situation, he recommends 'mutual coercion, mutually agreed upon by the majority of the people affected' (ibid., p. 1247). Later, Nobel Prize laureate in economics Elinor Ostrom has argued, in her acclaimed *Governing the Commons* from 1990, that the tragedy is not a *necessary* outcome of a common property situation. Ostrom documents several cases of successful management systems for shared resources, most of them of a local, traditional, unintended and spontaneously developed character. Her main argument is that, under certain conditions (we return to this in Chapter 2), local users of a relatively limited common-pool resource may agree among themselves on regulatory principles, making external intervention – as Hardin prescribed – superfluous, or even potentially harmful.

A few years before Ostrom's book, Bonnie McCay and James Acheson (1987) introduced the *co-management* literature with *The Question of the Commons*. Taking state involvement as a possibility (unlike Ostrom in her choice of empirical cases), they hold that the prospects for successful regulation of common property are improved if user groups can influence the formulation of rules and public management procedures (*co*-management rather than *self*-management). Compliance is not a main issue for Ostrom or the co-management writers – but legitimacy is. And at least implicit is Tyler's assumption that people are more likely to comply with rules that they consider legitimate – because the rules are 'good', because they as users have contributed to the production of the rules or influenced the process, or because they have trust in those who produced the rules. The literature on compliance in fisheries has tended to focus on the dichotomy between deterrence and legitimacy-induced compliance.

Compliance became an issue in the study of international relations (IR) largely as a reaction against the realist school that had dominated the field in the first decades after the Second World War. In the realist tradition, state compliance with international agreements was not an issue, because states were not believed to enter into agreements not in their interest. Moreover, for the realist, state behaviour is likely to conform to treaty rules because both the behaviour and the rules reflect the interests of powerful states. More recently, in line with the institutionalist criticism of the realist school that emerged in the late 1960s, the literature on compliance in IR has claimed that 'institutions matter'. The way an international agreement or an international regime is structured affects the propensity of states to comply. In their much-cited book *The New Sovereignty: Compliance with International Regulatory Agreements*, Abram and Antonia Chayes (1995) argue that transparency in the workings of regimes or treaties, mechanisms for dispute settlement and technical and financial assistance are all factors that enhance state compliance with international agreements. Moreover, decisions about compliance are not made once and for all. States have continuing relationships with each other over a range of issues, and questions of compliance arise in an environment of diffuse reciprocity. Negotiation does not end when a treaty is concluded: it is a continuous aspect of living under the agreement. Securing compliance with a treaty becomes a matter of 'bargaining in the shadow of the law' (Mnookin and Kornhauser, 1979; Cooter et al., 1982) or 'post-agreement bargaining' (Jönsson and Tallberg, 1998).

The various perspectives offered in the compliance literature emphasize the need for empirical investigation somewhat differently. The rational-choice approaches to the study of common-pool resource management – like those inspired by Hardin's 'tragedy of the commons' – place limited emphasis on empirical investigation, at least as regards testing that involves their basic assumptions. A shared resource is, for analytical purposes, assumed to be doomed to extinction unless state coercion is introduced. In its classical type, this is also the case with the realist perspective in IR: the most powerful state is believed to always have its way. Alternative perspectives take one step back, so to speak, and ask whether this is in fact the case. Can common-pool resources be successfully regulated by other means than strict state control? Are there other factors than a state's location in the inter-state hierarchy that can explain its behaviour? In the broader compliance literature, focusing on individuals' compliance with the law, this division is less clear, but it is still evident. The deterrence-oriented approach, which Tyler (2006) calls the instrumental perspective to compliance, in its purest form assumes that individuals always act out of self-interest in deciding whether or not to obey a law. By contrast, the normative perspective does not hold that individuals will necessarily base their behaviour on morality or legitimacy. It is open to the possibility that deterrence may be an important source of compliance, perhaps even the most important source, but it calls for a wider investigation of all possible sources of individual compliance with the law. So while the 'traditional' perspectives also welcome empirical analysis, the scope they prescribe is more limited. Instrumentalists may inquire into the workings of deterrence in various empirical settings – for instance with the aim of prescribing more effective deterrence. Likewise, rational theorists may study resource management with a view to finding the most suitable level and form of state coercion. Realists in IR *do* study world politics, but they explain events in terms of power relations. None of these scholars would question the basic assumptions of their respective research programme, for instance why actors behave the way they do. On the other hand, the 'alternative' approaches to common-pool resource management and compliance at individual and state levels all prescribe the empirical testing of all thinkable explanations of individual and state behaviour, including those preferred by the 'traditionalists'. This will be further elaborated in Chapter 2.[1]

THE EMPIRICAL SETTING

The Barents Sea (see Figure 1.1) lies north of Norway and north-western Russia, bounded in the north by the Svalbard archipelago and in the east by Novaya Zemlya. The rich fish resources of the Barents Sea have traditionally provided the basis for settlement along its shores, especially in northern Norway and the Arkhangelsk region of Russia. Since the 1917 Russian Revolution, the city of Murmansk on the Kola Peninsula has been the nerve centre of Russian fisheries in the Barents Sea.

The first steps towards international collaboration in managing the marine resources of the north-east Atlantic came as early as 1902, with the establishment of the International Council for the Exploration of the Sea (ICES). The European Overfishing Convention of 1946 introduced the first regulatory mechanisms in the form of minimum mesh size and minimum length of fish brought to land. In 1959, fourteen countries, among them Norway and the Soviet Union, signed the North-East Atlantic Fisheries Convention. The mandate of the North-East Atlantic

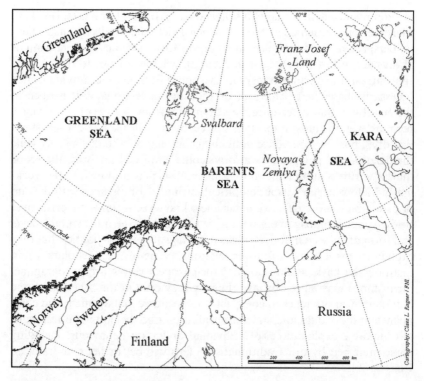

Figure 1.1　The Barents Sea

Fisheries Commission (NEAFC) was to provide recommendations on technical regulations, which could be done by simple majority, and fish quotas, which required a two-thirds majority. NEAFC did not succeed in introducing quotas until 1974–75. At the same time, agreement was reached on 200-mile EEZs at the Third United Nations Conference on the Law of the Sea, and Norway and the Soviet Union began negotiating bilateral management of Barents Sea fish stocks.

When Soviet Minister of Fisheries Aleksandr Ishkov visited Oslo in December 1974, the two countries agreed to establish a joint fisheries management arrangement for the Barents Sea.[2] The agreement was signed in Moscow in April 1975 and entered into force immediately. It is a framework agreement, in which the parties state their willingness to work together for the 'protection and rational use of marine living resources' in the NEAFC area. The agreement also established the Joint Norwegian–Soviet (now Russian) Fisheries Commission, which was to meet at least once every year, alternately on each party's territory. At the time, the details of the commission's work were not clear, but when the first session took place in January 1976 the parties had agreed to manage jointly the two most important fish stocks in the area, cod and haddock, sharing the quotas 50–50. In 1978, they agreed to treat capelin as a shared stock, and split the quota 60–40 in Norway's favour. When Norway and the Soviet Union declared their EEZs in January and March 1977, respectively, the bilateral cooperation agreement from 1975 was supplemented by a separate agreement on mutual fishing rights.[3]

During the 1980s, a specific quota exchange scheme developed between the parties, whereby the Soviet Union gave parts of its cod and haddock quotas in exchange for several other species found only in Norwegian waters. These species, especially blue whiting, were found in large quantities but were of little commercial interest to Norwegian fishers. In the Soviet planned economy, volume was more important than (export market) price, so the arrangement was indeed in the mutual interest of both parties.

This changed with the dissolution of the Soviet Union and the introduction of the market economy in Russia. Now cod and haddock, both high-price species on the global fish market, attracted the interest of not only Norwegian but also Russian fishers. Transfers of cod and haddock quota shares from Russia to Norway were reduced, and Russian fishing companies began to deliver their catches abroad, primarily in Norway. For the first time, Russian fishers had a real incentive for overfishing their quotas, while Russian enforcement authorities lost control of Russian catches, since quota control had traditionally been exercised at the point of delivery. Norwegian fishery authorities in 1992–93 suspected that the Russian fleet was overfishing its quota, and took steps to calculate total Russian catches, based on landings from Russian vessels in Norway and at-sea inspections by the Norwegian Coast Guard. Norway then claimed that Russia had overfished its quota by more than 50 per cent. The Russian side did not dispute this figure, and the two parties agreed to extend their fisheries collaboration to include enforcement as well. This involved exchange of catch data, notably the transfer by the Norwegian authorities to their Russian counterparts of data on Russian landings of fish in Norway. The successful establishment of enforcement collaboration was followed by extensive coordination of technical regulations, and joint introduction of new measures throughout the 1990s.

Around the turn of the millennium, a new landing pattern emerged. Russian fishing vessels resumed the old Soviet practice of delivering their catches to transport ships at sea. Instead of going to Murmansk with the fish, however, these transport vessels now headed for other European countries: Denmark, the UK, the Netherlands, Spain and Portugal.

Norway again took the initiative to assess the possibility of overfishing, but now encountered a less cooperative Russian stance. Thereupon Norway took unilateral measures to calculate overfishing in the Barents Sea, and presented figures that indicated Russian overfishing from 2002, rising to nearly 75 per cent of the total Russian quota in 2005, gradually declining to zero in 2009. The Russian side never accepted these figures, claiming they were deficient at best, and an expression of anti-Russian sentiments at worst. ICES, however, used them in its estimates of total catches in the Barents Sea during the 2000s, thereby providing these figures with some level of approval. Other issues of contention were disagreement about quota levels around the turn of the millennium and (to a lesser extent) about methods for estimating stocks in the mid-2000s.

QUESTIONS TO BE ASKED

Since the dissolution of the Soviet Union, claims of overfishing in the Barents Sea have come largely from the Norwegian side and targeted Russian fishers.[4] The Norwegians were not content merely to document the state of affairs and let the Russians take care of any problems that might exist. Instead, they engaged in active bargaining to induce Russian fishers to comply with fisheries regulations in the Barents Sea, and to persuade the Russian state authorities to take seriously their commitment to sustainable fisheries management – as a treaty partner in the Barents Sea, and as signatory to all the main global fisheries agreements. In this book we will look more closely into these negotiation efforts, employing the theoretical framework of post-agreement bargaining.

The focus is on the bilateral management arrangement as a communications channel for negotiating the practical terms for good fisheries management according to the bilateral agreements between Norway and Russia, as well as the global fisheries agreements the two countries have adopted. The question is not to what extent the Norwegian–Russian fisheries management regime has been effective (in solving a problem, improving a situation or generally making any difference).[5] Nor do I examine it as a location for spontaneous learning for the actors involved. No, I focus on Norway's deliberate efforts to influence Russian behaviour in a direction it considers to be in compliance with national law (regarding the behaviour of individual fishers) and international obligations (for Russia as a state). This involves asking the following questions:[6] What form have the Norwegian negotiation efforts taken? How have these efforts been perceived by the Russian actors? Is there an indication that the efforts have had any effects on Russian behaviour?

At the individual level, the focus is on encounters at sea between inspectors of the Norwegian Coast Guard and the captains of Russian fishing vessels, usually during inspections. How do the inspectors try to induce the captains to comply with regulations?[7] How do the captains perceive the Norwegian inspections? Is there reason to believe that the Norwegian inspectors make a difference for compliance among Russian fishers? If so, does this happen through deterrence, or by influencing the captains' ethical reasoning, or the legitimacy of the rules or of the management system? At state level, I examine how the Norwegian fishery authorities, through the Joint Fisheries Commission and its Permanent Committee (see Chapter 3), as well as in day-to-day communication with their Russian counterparts, attempt to influence Russian compliance with the bilateral fishery agreements and global fishery treaties.[8] Focus is on the two instances when Norway accused Russia of overfishing, in the early 1990s and the mid-2000s. I also discuss Norway's attempts (i) to involve Russia in harmonizing a range of technical regulations between the two countries, and jointly introduce new regulations in the second half of the 1990s, (ii) to persuade Russia to keep quotas as close as possible to ICES's scientific recommendations in the first years after the turn of the millennium, and (iii) to prevent new methods for estimating fish stocks, proposed by the leading Russian federal fisheries research institute in the mid-2000s (but not considered by ICES to be precautionary), from being officially adopted by the Russian fishery authorities. To the extent that Russia has complied with obligations to conduct sustainable fisheries management – which, to a large extent, it actually has in the Barents Sea, later if not sooner – can this be attributed to Norway's endeavours to influence Russian practice?

In asking these questions, I place myself in the 'alternative' perspective on compliance. As noted above, the 'traditional' approaches are less inclined to encourage empirical investigation of why subjects comply, or how the public authorities best can ensure compliance. People are believed to obey the law because they see it in their best economic interest, and the authorities provide surveillance and sanctions to ensure deterrence. States, in turn, are seen as behaving according to the will of more powerful states, whether prescriptions are encoded in international agreements or not. The 'alternative' perspectives do not deny that this might be the case, but they want to see it empirically tested. They also point out that empirical testing has shown the world to be different from that presented in the rationalist models. People do comply, even in the absence of economic self-interest or threat of punishment. Users of a common-pool resource do ensure successful management even without state coercion. State compliance is determined by institutional aspects of the treaty or regime, or by post-agreement communicative action.

The focus here is on the communicative aspects of the 'alternative' perspective, but I take the role of legitimacy and morality into account in assessing what makes Russian fishers follow the guidelines of Norwegian inspectors (to the extent that they do). Likewise, institutional structure is relevant when I discuss why Russia complies with Norwegian requests under the bilateral fisheries management cooperation arrangement (to the extent that Russia does so). When bargaining is successful, is that a result of the bargaining itself (does it, for instance, perhaps take a particularly suitable form)? Or would the subject (Russian fishers, or Russia as a state) see compliance as the preferred alternative anyway? Would it be in their long-term, if not short-term, interest? Have the Russians been convinced by the Norwegians' substantive arguments? Or is the decisive factor that the negotiations have taken place in this particular institutional setting?

Finally, a note on the terms 'compliance' and 'post-agreement bargaining'. In line with Young's definition (1979, p. 4), I understand compliance to mean 'all behavior by subjects or actors that conforms to the requirements of behavioral prescriptions or compliance systems' (see Chapter 2 for more on these concepts). In discussing individual compliance here, it is fairly simple to state what constitutes compliance: behaviour that conforms to the established fisheries regulations in the Barents Sea. At the state level, however, things are less straightforward. Our point of departure is the bilateral fisheries agreements between Norway and Russia, negotiated and signed in the mid-1970s. However, these are framework agreements that began with little prescriptive substance and have continuously been filled in at the annual sessions of the Joint Commission. Hence, what I analyse is whether the parties have complied with the measures fixed in the protocols from these sessions. This also involves drawing into the discussion the parties' obligations according to the law of the sea. To a large degree, the obligations emanating from these various levels merge into the overarching demand to conduct sustainable fisheries management, and, since the mid-1990s, fisheries management that is in accordance with the precautionary approach (see Chapter 4). The discussion will not be limited strictly to compliance bargaining, which can be understood as a sub-group of post-agreement bargaining. Jönsson and Tallberg (1998, p. 372) define post-agreement bargaining as 'all those bargaining processes which follow from the conclusion of an agreement', as we shall see in Chapter 2. They understand compliance bargaining as 'a process of bargaining between the signatories to an agreement already concluded, or between the signatories and the international institution governing the agreement, which pertains to the terms and obligations of this agreement' (ibid.). As noted, in discussing the state level I include Norway's efforts to influence Russian views on scientific recommendations and technical regulation. That point is normally seen as related not to the law-abidingness of a fishery, but to fisheries management more widely. It is a matter of interpretation whether this can still be understood as 'pertain[ing] to the terms and obligations of [the] agreement'. To indicate the slightly broader approach taken here, I generally employ the concept of 'post-agreement bargaining' rather than 'compliance bargaining'. On the other hand, following the standards set for precautionary fisheries management by ICES and the Joint Commission is also a matter of compliance with international fisheries law, as we will see in Chapter 4.

METHODOLOGICAL CONSIDERATIONS

The study builds on various qualitative research methods: observation, interviews and textual analysis. Fresh empirical data were collected through interviews with Russian fishers (most of them captains) in 2009–10. The interviews were conducted by myself and my colleague Anne-Kristin Jørgensen (some by Anne-Kristin alone) in various Norwegian ports[9] where the Russian fishing vessels had come to deliver fish or have repairs done. We would get in touch with some of the captains through their Norwegian agents (who handle practical matters for them while they are in Norwegian port) or mutual acquaintances; some we would contact directly – on the docks, in the streets or by simply climbing on board their vessels. We had conducted similar interviews with Russian captains in Norwegian ports in 1997–98; these interviews also are used here.[10] It proved far more difficult to get the Russian captains to talk to us this time than in the late 1990s. In fact, most of those we approached refused to speak with us, saying that the shipowner had forbidden them 'to talk with journalists'. In order to get a good amount of interviews, we therefore instigated a separate round of interviews, to be carried out by a Russian researcher in Murmansk.[11] She sought out captains and other crew members on fishing boats from acquaintances and business contacts in the city's fishing industry. All in all, around fifty people in this category were interviewed. Approximately the same number of scientists and civil servants (including representatives of Russian fisheries enforcement bodies) were consulted during the period 2006–09.[12] As follows from the description of how we found our interviewees, the sample was not randomized. That would not have been practically possible; nor is randomization a requirement in qualitative interviewing. A general guideline in qualitative research is to continue interviewing until you reach theoretical saturation (Glaser and Strauss, 1967) – that is, until each new interview does not add any new information to what you already have. In our interviews in the late 1990s, we felt that we reached saturation point rather soon; most of our Russian respondents produced quite similar stories. A decade later, practical obstacles restricted the number of interviews. The stories that we heard differed more, although certain trends were discernible. A typical interview lasted for about an hour; some went on for a couple of hours. Interviews were semi-structured and open-ended. The objective was to acquire as good an understanding as possible of the interviewees' experiences (see limitations in this aim below) instead of merely collecting factual information. Hence, we encouraged respondents to speak at length on topics that spurred their interest. Since both my co-interviewer and I speak Russian, all interviews were conducted without interpreter.[13]

In addition, I build on (partly participant) observation in the Norwegian Coast Guard, and in the Joint Norwegian–Russian Fisheries Commission and its Permanent Committee. I acquired my 'cultural competence' (Neumann, 2008) about the Barents Sea fisheries while engaged as Russian interpreter and fishery inspector in the Norwegian Coast Guard from 1988 to 1993.[14] My presentation of the encounters between Norwegian inspectors and Russian fishers in Chapter 5 builds largely on observations from this engagement.[15] After leaving the Coast Guard, I continued to work up till 2000 as an interpreter for the Norwegian fishery authorities. I participated regularly in the Joint Commission's Permanent Committee as well as at joint Norwegian–Russian seminars for fishery inspectors, and on occasion in the Commission itself. In the mid-2000s, I was engaged to write an anniversary publication for the Joint Commission's thirtieth anniversary in 2006, and attended sessions in the Commission as (non-participant) observer formally included in the Norwegian delegation. On these occasions, I was free to report my observations, except from internal meetings of the Norwegian delegation. As to my participant observation, I have used my best judgement in choosing what to report, following the general guideline of not reporting incidents or practices that have not been – or at least could not have been – referred to in the media or in other public form. Finally, I use protocols from the sessions of the Joint Commission and its Permanent Committee, as well as articles from Norwegian and Russian media.[16]

As to the research questions outlined above, I use mainly my own observations, combined with written materials, to describe Norway's efforts at negotiating Russian compliance. In part I employ observation and textual analysis in describing Russian perceptions of these efforts, too, but here my interviews serve as the main source. All this is fairly straightforward in methodological terms, although care must be taken to ensure thorough interpretation and avoid unfounded generalization. More contested is the possibility of saying something about *why* actors actually comply. I can report what captains of Russian fishing vessels say in my interviews with them, but cannot really know what goes on in their heads when they make decisions about compliance. I can describe Russian political action in the country's fisheries relations with Norway, but I cannot state exactly why a particular choice of action was made. Implicitly, I cannot prove causal relations between Norwegian negotiation efforts and Russian 'response'. Criticism of the tendency to focus on actors' motives has come from the *narrative turn* in sociology (see, for instance, Gubrium and Holstein, 2009) and other social sciences, including IR (see, for instance, Ringmar, 1996). Since we cannot say anything about the true motives of actors, we should focus instead on how they frame their arguments and how this influences the political choices that are available to other actors.[17] I do not claim to know my interviewees' genuine experiences or motives – but this book is not intended as a narrative analysis.[18] In scrutinizing the statements made by my interviewees – realizing that their statements could, for instance, reflect the dominant stories in the interviewees' community or what they expect the interviewer wants to hear, more than their 'genuine experiences'[19] – I follow the tradition in the compliance literature of treating as relevant what people say about their behaviour and perceptions. We can never know exactly what motivates individuals: what they say is the best indication we can get. On a more general note, I follow the qualitative research standard of triangulation, viewing information from different sources – texts, interviews and observation – in relation to each other.

Furthermore, there is the 'big question' – with both theoretical and methodological connotations – of what determines the foreign policy of a state. Since Graham T. Allison's classic 1971 study of the Cuban missile crisis (re-issued as Allison and Zelikow, 1999), the internal political processes of a state have received due attention in foreign policy analysis.[20] In this study, it would be too facile to ascribe Russian compliance with a Norwegian request to Norway's negotiation efforts. Whether related to these efforts or not, Russian foreign policy can be interpreted as the result of Russia assessing its best interests and behaving accordingly, or the more 'accidental' result of political bargaining or decision-making procedures in the Russian political system. Norwegian negotiation efforts can, in principle, have had an influence on Russia's calculated interest equation at the state level, on the perceived interests of specific groups in Russia and on Russian bureaucratic procedures as well. We will explore these alternative explanations, although the book relates mainly to the compliance literature and not the wider foreign policy tradition. In methodology, I follow the rather pragmatic approach (using different methods and different theoretical perspectives) often applied in the literature on the interface between domestic politics and state behaviour in international regimes, such as in implementation studies (see, for instance, Underdal, 2000).

THE BOOK

The theoretical and empirical views provided in this introductory chapter are fleshed out in more detail in the next two chapters. Chapter 2 provides a broader overview of the literature on compliance, the management of common-pool resources, compliance in fisheries and states' compliance with their international commitments. While searching for similarities in these various bodies of literature, I also indicate differences in substance and scientific ambition. Chapter 3 outlines the fish resources and jurisdiction of the Barents Sea, and the general traits of the bilateral fisheries management regime in the area, as well as the national fisheries management policy of Norway and Russia. Chapters 4 and 5 go on to describe and analyse Norwegian negotiation efforts and Russian responses at the state and individual levels. I give a chronological account of relevant events since the dissolution of the Soviet Union, starting in Chapter 4 with Norwegian allegations of Russian overfishing in the early 1990s, followed by various further attempts on the part of Norway to persuade Russia to harmonize technical regulations and jointly introduce new measures in the late 1990s, to get the country to follow scientific recommendations about quota levels in the years around the turn of the millennium, and most recently to reduce overfishing and refrain from introducing new methods for the assessment of fish stocks in the 2000s. Then in Chapter 5 I focus on how the Norwegian Coast Guard has tried to talk Russian fishers into compliant behaviour throughout the period, and how this has been perceived by the Russian side. Chapter 6 presents a summary of the findings, analysing them in relation to my theoretical points of departure. Above all, I ask whether we can draw some common lessons from my empirical discussion of post-agreement bargaining and compliance at the state and the individual levels.

NOTES

1. The terms 'alternative' and 'traditional' are not quite precise, which will also follow from the more thorough presentation of these theories in Chapter 2.
2. 'Avtale mellom Regjeringen i Unionen av Sovjetiske Sosialistiske Republikker og Regjeringen i Kongeriket Norge om samarbeid innen fiskerinæringen', in *Overenskomster med fremmede stater*, Oslo: Ministry of Foreign Affairs, 1975, pp. 546–9.

3. 'Avtale mellom Regjeringen i Kongeriket Norge og Regjeringen i Unionen av Sovjetiske Sosialistiske Republikker om gjensidige fiskeriforbindelser', in *Overenskomster med fremmede stater*, Oslo: Ministry of Foreign Affairs, 1977, pp. 974–8.
4. This is not to say that Norwegians have never overfished their quotas in the Barents Sea. Until well into the 1980s (see Chapter 3), Norway was in fact allowed to – and regularly did – overfish its total quota with traditional passive gear. During the 2000s, some (limited) Norwegian overfishing took place, since the authorities could not halt the fishery for a specific group of vessels fishing on a 'competitive' basis once the total quota had been taken. Nor is this to say that the Russians actually *did* overfish their quotas in the early 1990s and during the 2000s (or that they did *not* do so in the late 1990s). Assessing Norway's claims of Russian overfishing is beyond the scope of this book.
5. Stokke (2010a, forthcoming) is the authoritative source here.
6. Again, this is not to say that there is an *a priori* reason to assume that Russia should learn from Norway, on whatever empirical or moral grounds. I simply start from the empirical observation that the Norwegians have sought to influence Russian behaviour in their dealings with Russia in the Joint Commission and in encounters between Norwegian inspectors and Russian fishers at sea.
7. The Norwegian Coast Guard enforces Norwegian fishery regulations in areas under Norwegian jurisdiction in the Barents Sea (see Chapter 3). Indirectly, the Coast Guard also contributes to the enforcement of Russian regulations. Many technical regulations have been harmonized between the two countries (see Chapters 3 and 4). And when the Norwegian Coast Guard checks the amount of fish on board a Russian fishing vessel at the time of inspection, this also provides information that can be used by the Russian enforcement authorities in their quota control. The Norwegian Coast Guard can charge a Russian vessel for underreporting of catch if more fish is discovered on board than documented in the catch log. Formal charges of *overfishing*, i.e. of taking more fish in the course of a year than one is entitled to, can be made only by the national authorities of the vessels in question, in this case the Russian authorities.
8. Again, this is not to imply that the Norwegian position is necessarily more 'correct' than the established Russian practice that Norway tries to influence. My empirical focus is on the behaviour of Norway, for two reasons: First, most proposals for change in the bilateral management regime after the dissolution of the Soviet Union have come from Norway. Second, empirical data on state policies are more readily available from the Norwegian side than from Russia.
9. To ensure the anonymity of our Russian interviewees, the exact locations are not specified.
10. These interviews were conducted for my Ph.D. dissertation, which was published in revised form as Hønneland (2000a).
11. We decided against going to Murmansk to interview Russian fishers there ourselves, since Norwegian allegations of Russian overfishing had become a very sensitive issue in north-western Russia at the time. In order to get a proper Russian visa, we would have had to state the aim of our research and name the institutions we planned to visit. There is a chance that we would not even have been granted a visa to interview members of the Russian fishing industry about possible overfishing. If we had acquired a visa, it might have been difficult to preserve the anonymity of the interviewees, since representatives of security services (or other Russian authorities) might follow us. Given the sensitivity of the issue, we do not identify the Russian researcher who conducted the interviews either.
12. This figure includes interviews I conducted during this period for other research projects and consultancies of relevance to this book – in particular the interviews carried out for my anniversary publication on the Joint Norwegian–Russian Fisheries Commission to its thirtieth anniversary (Hønneland, 2006) and those that I conducted together with Bente Aasjord from Bodø University College on knowledge disputes in Russian fisheries science (Aasjord and Hønneland, 2008). Prior to that, I had

regularly interviewed Russian civil servants for a monograph about Russian fisheries management (Hønneland, 2004).

13.   We did not use tape recorders in the interviews, but took care to note down what was said as accurately as possible and compare our notes immediately after the interview. For the pros and cons of using a tape recorder, see Rubin and Rubin (2005, pp. 110–12). Among students of Russian politics and society, there is a clear tradition of not using a tape recorder, so as not to intimidate interviewees; see, for instance, Ries (1997, p. 6).

14.   I started as interpreter and was also used as a witness at fishery inspections. Then I was trained as a fishery inspector, and for the two last years of my engagement I was allowed to conduct inspections on my own.

15.   I lean on my own reports from the time, submitted to my superiors on shore after each trip.

16.   I have chosen a 'medium-level' reference style. Since my primary aim is not documentation, as for instance a historical or legal text would be – my aim is more in the field of theoretical discussion – I do not provide the specific source for every single event that is presented. However, I indicate the source of direct citations – and of course any fact that I build on secondary sources. Apart from that, I state the main source used in the different sections of the book, like observation, interviews or protocols from the Joint Commission. The protocols are published in Norwegian and Russian; English translations are mine. Primary sources are cited in the chapter notes, not in the reference list.

17.   See, for instance, Krebs and Jackson (2007, p. 42): 'We cannot observe directly what people think, but we can observe what they say and how they respond to claims and counter-claims. In our view, it does not matter whether actors believe what they say, whether they are motivated by crass material interest or sincere commitment. What is important is that they can be rhetorically maneuvered into a corner, trapped into publicly endorsing positions they may, or may not, find anathema.' See Jackson (2010) for a more thorough discussion.

18.   However, I myself am familiar with and positively disposed to narrative analysis. In Hønneland (2010), I investigate the narrative practices of Kola Peninsula residents and the room for manoeuvre these practices leave for foreign policy action in the region.

19.   A major claim of narrative analysts is that people draw on a limited reservoir of narrative resources when they tell stories. These resources, which vary with time and across space, do not only reflect who people are: they *make* them who they are. See, for instance, Somers (1994) and Gergen (2001). Similarly, the literature on qualitative interviews is full of guidelines for interpreting how respondents talk. Rubin and Rubin (1995), for instance, single out various mediation forms by which information can be conveyed from interviewee to interviewer: narratives, stories, myths, accounts, fronts and themes. I will explain these concepts in Chapter 5.

20.   Briefly stated: Allison and Zelikow (1999) claim that, besides treating a state as a 'rational unitary actor', researchers should analyse a state's foreign policy as the result of bureaucratic processes and negotiations between and among interest groups (or individuals) within the state.

2. Common-pool resource management and compliance with international commitments

The investigation here is narrowly focused in one sense, and extensive in another. Empirically, I use only one case study, but include practices at the individual and state levels in the discussion. As regards theory, I draw on several distinctly different bodies of literature, which communicate with each other only to a limited extent. This chapter gives an overview of these theory traditions: the general study of (individual) compliance, research on the management of common-pool resources, the literature on (individual) compliance in fisheries, approaches to state compliance with international law, and (state-level) post-agreement bargaining. As we will see, these separate bodies of literature share a number of similarities.

THE STUDY OF COMPLIANCE

Compliance and law enforcement have been studied mainly within the fields of economics, criminology, psychology and sociology.[1] In the economics literature, the theme is found as early as in the work of Adam Smith (1759, 1776), who noted that individuals who act in the pursuit of self-interest may impose harm on others and must therefore be restricted in some way. He also made the link between crime and economic circumstances, claiming that individuals most often resort to criminal activity when their opportunities for a lawful income are slim. Jeremy Bentham (1789), following Smith, argued that criminal behaviour is economically rational and deterrence necessary to reduce crime. The early 1900s saw numerous attempts to explore the link between crime and economic circumstances (see e.g. Bonger, 1916). Most of these studies found that changes in the economic situation of the working class, as well as growing disparities in wealth between classes, were associated with changes in the level of crime. A formal theoretical framework for explaining criminal activity was developed in the late 1960s. Following Smith and Bentham, Gary Becker (1968) argued that criminals behave basically like all other individuals when they attempt to maximize personal utility. According to the model, an individual commits a crime if the expected utility from doing so exceeds the utility from engaging in lawful activities. Becker's contribution inspired a series of studies on the economics of crime (see, for instance, Pyle (1983) for an overview).

By contrast, research in psychology and sociology has emphasized the importance of socialization, morality and legitimacy in eliciting compliant behaviour, building on the Durkheimian view that the law is an expression of collective morality, and that the primary function of punishment is to restore the normative order that was challenged by the violator. The symbolic function of enforcement is emphasized; that is, enforcement can have an effect beyond deterrence. Gray and Scholz (1993), for instance, found that enforcement did not only (or not necessarily) lead to a reduction in violations, but led also to greater attention to the problem that enforcement sought to solve. In their study of compliance in business corporations, Kagan and Scholz (1984) claim that there are three popularly held images of non-compliance in that sector: business firms as amoral calculators (where non-compliance stems from economic calculation), as political citizens (where at least some non-compliance follows from the fact that subjects view regulations or orders as arbitrary or unreasonable), and as organizationally incompetent entities (where managers fail to maintain a high level of compliance owing to institutional default). If regulated businesses are seen as amoral calculators, the regulatory agency should emphasize aggressive inspection of all firms and promptly impose severe legal penalties for all violations. The goal is deterrence, and the inspector takes the role of the strict *policeman*. If businesses are viewed as political citizens, on the other hand, the inspector should act more as a *politician* who tries to persuade people of the rationality of the regulation in question: 'But he also should be willing to suspend enforcement, to compromise, to seek amendments to the regulations. In short, he should be responsive to the "citizen's" complaints, ready to *adapt the law* to legitimate business problems created by strict enforcement' (ibid., p. 68, emphasis in original). Finally, if businesses are thought to be prone to incompetence and regulatory violations owing to organizational failure, the inspector should serve as a *consultant*, aiming to disclose knowledge gaps within the organization and educate employees to improve the situation.

Oran R. Young is one of the few writers who have combined perspectives on individual and state compliance. As early as 1979, he published a (largely theoretical) monograph on compliance at both these levels, which incorporated aspects of both instrumental and normative perspectives. In this work, Young (1979, pp. 2–6) defined compliance as all behaviour by subjects that conforms to the requirements of behavioural prescriptions within a specific compliance system. A behavioural prescription is 'any well-defined standard setting forth actions (including prohibitions) that members of some specified subject group are expected to perform under appropriate circumstances' (ibid., p. 2) – in other words, a rule that provides guidelines for certain conduct and actions. A compliance system is a set of behavioural prescriptions designed to regulate an interdependent group of human activities in a coherent fashion. Further, the subjects of a behavioural system are the units that must ultimately choose whether or not to comply with any given prescription. In this context, a 'subject' is any entity that possesses preferences concerning alternative states of the world and is capable of engaging in choice behaviour. The factors that in some way or other influence these preferences are called 'bases of compliance'.

Young (1979, pp. 18–25) proposes six such bases of compliance. First, there is self-interest. Compliance may emerge as the preferred option even in the absence of external influence if subjects simply conclude that the expected value of compliance outweighs that of violation. Second, enforcement involves explicit attempts by the authorities to manipulate the cost–benefit calculations of subjects, primarily through the threat of sanctions in the event of detected violation. Third, inducement entails an attempt to raise the expected value of compliance rather than to reduce the expected value of violation; this normally involves rewards of some kind. Fourth, subjects' actions may be shaped by social pressure stemming from other external actors than public authorities. Fifth, behaviour may be influenced by a sense of obligation (which to various extents is externally induced). Sixth, decisions by subjects concerning compliance versus violation are often influenced by subconscious or unconscious considerations; the terms 'habit' and 'practice' here refer to patterns of behaviour that are acquired by frequent repetition. Finally, Young recognizes that individuals seldom make decisions independently of the behaviour of others. Rather than singling out this as a separate base of compliance, however, he holds that the behaviour of other people has an impact on how each of the other bases of compliance affects a subject's choice situation.

TWO TRADITIONS IN THE STUDY OF COMMON-POOL RESOURCES

A common-pool resource can be understood as a natural (or in some instances man-made) resource sufficiently large to make it costly to exclude users from obtaining subtractable resource-units. Hence, the two criteria used to define a common-pool resource are (i) the cost of achieving physical exclusion from the resource, and (ii) the presence of subtractable resource-units (Gardner et al., 1990). The second criterion entails that acquisition of resource-units by one user occurs at the expense of other potential users. This is the case with fish and with grazing lands, for instance, but not with, say, weather forecasts or traffic lights. Common-pool resources are distinct from public goods because subtractability is higher, and from private goods to the extent that exclusion is more difficult (Ostrom et al., 1994).

The exploitation and management of common-pool resources have attracted the interest of economists, political scientists, sociologists and anthropologists, among others. The focal question is how the social setting can be organized to best enable continued renewal of the resource. Garrett Hardin (1968) recommended 'mutual coercion, mutually agreed upon'. The idea that common property causes intricate collective dilemmas is not new. Aristotle recognized this more than two thousand years ago, commenting that 'what is common to the greatest number has the least care bestowed upon it. Everyone thinks chiefly of his own, hardly at all of the common interest' (cited in Ostrom, 1990, p. 2). Hobbes's parable of man in a state of nature is the prototype of the tragedy of the commons: humans seek their own good and end up fighting one another because of resource scarcity. This state of war of everyone against everyone can be avoided only through a social contract in which individuals concede authority to a sovereign, agreeing to respect legal restrictions on their relations to each other and the sovereign's right to enforce those restrictions.[2] In modern times – and more than a decade before Hardin – Scott Gordon (1954) expressed a similar logic in another classic, 'The Economic Theory of a Common-Property Resource: The Fishery'. Here he argues that the rational action for each individual in an open-access fishery is to take as much catch as possible before others appropriate it. The result is increased fishing effort in terms of labour and capital, and a situation in which the costs of production exceed rent or revenue. Again the argument is that, when a resource is common property, this invariably leads to its depletion.

Hardin's model has often been formalized as a *prisoner's dilemma* game (see, for instance, Dawes, 1975). In games of this sort, where there is no communication between the players, and they possess complete information on the pay-offs of various outcomes, each player is always better off opting to defect, no matter what the other one chooses. However, individually rational strategies lead to collectively irrational outcomes. A related view of the difficulty in getting individuals to pursue their common welfare was developed by Mancur Olson (1965) in *The Logic of Collective Action*. His argument rests largely on the assumption that an individual who cannot be excluded from the benefits of a common good once the good is produced will have little incentive to contribute voluntarily to supplying that good. (Unlike Hardin, Olson considers it an open question whether collective benefits will eventually materialize.) Olson counts as one of the classics in public choice theory. Using concepts initially developed in institutional economics, public choice theorists mainly see the management of natural resources as the determination of least costly alternatives (Sproule-Jones, 1982). Particular effort has been made to elaborate models based on state and on private ownership.

While the attraction of public choice theories on the management of natural resources lies in their ability to capture important aspects of real-life problems in a comprehensible manner, the danger, critics argue, is that the constraints defined for the purpose of analysis are taken as necessary theoretical presumptions or even as empirical facts. Most of the literature critical to these models focuses on attempts to prove these assumptions and prescriptions empirically wrong. Above all, critical studies point out that the tragedy is not a *necessary* outcome in a situation where resources are used in common. A great many successful management systems for common-pool resources have been documented (see, for instance, Baland and Platteau, 1996; Berkes, 1989; Bromley, 1992; McCay and Acheson, 1987; Ostrom, 1990; Ostrom et al., 1994; Pinkerton, 1989). Ostrom (1990, pp. 185ff.) proposes a set of criteria that should be met to create favourable conditions for the establishment of sustainable management systems. These criteria state, among other things, that the number of participants using the resource should be limited; that the rules for utilization should at least in part be designed by the users themselves; that monitoring of rules should be carried out by individuals accountable to the user group; and that sanctions should be used in the form of graduated punishment, implying that only severe and repeated violations should be given severe sanctions.

In anthropology, the workings of informal regulation systems have dominated fisheries studies. Such studies normally argue that small communities are better off in their pursuit of regulating a commons if there is no interference from state authorities. Traditional knowledge, local commitment and social norms, it is argued, are in many cases sufficient for successful management. Axelrod in 1984 (re-issued in 2006) showed how game theory can explain how stable cooperation patterns among individuals with conflicting interests can develop without the intervention of an external authority. In his study of repeated games with the reward matrix of a prisoner's dilemma game, a 'tit-for-tat' strategy predominated, in which participants started out with a readiness to cooperate, and defected only when their opponent did so first. Spontaneously developed regulation systems often function on the basis of sanction mechanisms similar to this tit-for-tat approach, for instance when fishers have their fishing gear cut off when they enter areas that other fishers consider 'theirs'. Acheson's (1975) study of the lobster industry off the coast of Maine is a classic in this respect. Studies such as Ostrom et al. (1994) and Baland and Platteau (1996) combine game theory and field studies in seeking to show how the tragedy of the commons perspective fails in explaining many real-life situations.

Finally, the co-management literature (see Chapter 1) has been a main arena for intellectual craftsmanship in social science studies of fisheries management since the mid-1980s (see, in particular, Jentoft, 1985, 1989, 2005; Jentoft and McCay, 1995; Jentoft et al., 2009; McCay and Acheson, 1987; McCay and Jentoft, 1996; Pinkerton, 1989; Sen and Nielsen, 1996; Wilson et al., 2003). Like representatives of the 'self-management' perspective (such as Ostrom), co-management theorists claim it is possible to create regulatory systems without recourse to coercion as the sole (or even most) feasible mechanism for securing compliance and thus avoiding the tragedy of the commons. For various reasons, they accept a certain degree of state involvement, for example in the form of a general responsibility for the management policy vested with central government. Nevertheless, they claim that the prospects for a successful regulatory system are significantly enhanced if user groups are allowed to influence the design of rules and management procedures. This can entail the delegation of power to user groups within specific functional fields or geographical areas, the participation of user-group representatives in bodies where regulations are formulated, or the acceptance and formalization by state authorities of traditional regulation mechanisms based on social norms within the community that uses the commons.

COMPLIANCE IN FISHERIES

Compliance became an explicit issue in the literature on fisheries management with the work of Jon G. Sutinen (see, for instance, King and Sutinen, 2010; Kuperan and Sutinen, 1998; Sutinen and Andersen, 1985; Sutinen and Kuperan, 1999; Sutinen et al., 1990).[3] While compliance had been an implicit issue in earlier work on fisheries economics, such as Gordon's (1954) (see above), Sutinen brought to the forefront of empirical analysis the fishers' inclination to comply with the law. Moreover, while drawing on the classical deterrence literature, he also brought normative issues into his theory framework, which in a recent paper he terms the 'enriched theory of compliance' (King and Sutinen, 2010, p. 8). This theory takes into account the individual's personal morality, the perceived fairness of rules and procedures, and peer pressure, in addition to deterrence. While the degree of the influence of each of these factors varies from case to case, it is claimed that '[i]n most fisheries normative influences result in most fishers complying with fishing restrictions despite potential economic gains from doing otherwise' (ibid., p. 8).

Sutinen's work has influenced most subsequent analysis of compliance in fisheries. First, economists have specified his theory in formal economic models (Hatcher and Gordon, 2005; Hatcher et al., 2000; Nøstbakken, 2008). Second, social scientists and practitioners have produced more elaborate practical approaches to the study of compliance in fisheries, based on Sutinen's work. Randall (2004, p. 293) holds that the following determinants of fisher behaviour should be taken into account: instrumental influences (economic yield and effectiveness of enforcement), normative influences (legitimacy of the management regime and fairness of outcomes) and social influences (behaviour of others and personal morals). Nielsen (2003, p. 431) sketches a model with the following main components: industry structure, control and enforcement, as well as internal obligations (including legitimacy of procedure and outcome, and social and individual norms). I have myself produced a model that explicates various coercive and discursive compliance mechanisms that public authorities can apply to induce compliance in fisheries, including enforcement (coercive measures) and attempts to influence fishers' habits, sense of obligation or conviction (discursive measures), in the various subsystems of compliance (research, regulation and control) (Hønneland, 1999a). Finally, the theoretical framework underlying these contributions has spurred several case studies of compliance in fisheries, including in the USA (King and Sutinen, 2010; King et al., 2009; Randall, 2004), the UK (Hatcher and Gordon, 2005; Hatcher et al., 2000), Canada (Gezelius, 2003, 2004), Australia (Bose and Crees-Morris, 2009), South Africa (Hauck, 2008; Hauck and Kroese, 2006), Indonesia (Crawford et al., 2004), Denmark (Nielsen and Mathiesen, 2003), Norway (Gezelius, 2002, 2003, 2004, 2006), Sweden (Eggert and Ellegård, 2003) and Norway/ Russia in the Barents Sea (Hønneland, 1998, 1999b, 1999c, 2000a, 2000b). A common theme is, as indicated by King and Sutinen (2010), that at least some of the compliance observable among fishers can be explained by normative factors – that is, other influences than deterrence.

The work of Stig S. Gezelius on the relationship between norms, enforcement and compliance deserves particular mention. He holds that theory developments in the study of compliance in fisheries have been few, and that research has focused almost exclusively on the extent to which various factors influence fishers' choices, while in-depth studies of how and under what conditions the different factors affect their choices are lacking (Gezelius, 2002, p. 306). Among his findings from ethnographic research in fishing communities in Norway and Canada is that social control is an important deterrent in fisheries, besides formal enforcement. It is especially violations that other fishers perceive as an expression of greed that give rise to social sanctions, while violations committed in order to secure a satisfactory living are tolerated to a far greater extent. However, he found that formal enforcement was necessary for informal social control; violators were subjected to informal sanctions only if the violated regulation was formally enforced (Gezelius, 2003, 2004, 2007). As regards theory, he calls for greater attention to what he terms the Durkheimian enforcement mechanism, which emphasizes the symbolic meaning of enforcement (like the effect of formal enforcement on social control), as opposed to the Hobbesian and Habermasian mechanisms, with their emphasis on deterrence and rational communication, respectively (Gezelius, 2007). Further, from his fieldwork he noted a difference in how legitimacy determined compliance in Norway and in Canada. In Norway, law-abidingness was in itself a social norm; it formed part of the collective image of the *bona fide* fisher. In his Canadian case, which was a small fishing community in Newfoundland, the legislators lacked such authority. Fishers regarded the law as a servant, not as a source, of morality. They found it reasonable to comply with regulations only if these were clearly aimed at restoring fish stocks, not simply because they were 'the law' (Gezelius, 2003, 2004, 2007). On the relationship between norms and outlooks for economic gain, he notes:

> Social norms and social control partly dissolve the connection between expected benefit and the likelihood of infractions. . . . However, . . . actors may pay regard to and take advantage of moral norms in a strategic and goal-oriented manner. The model of the brain-dead conformist is, in other words, as insufficient as the model of the atomistic, utility-maximizing opportunist. . . . [U]nderlying . . . fishermen's strategic adaptations and moral thoughtfulness is a deeply rooted sense of belonging which can hardly be ascribed reason alone. (Gezelius, 2002, p. 313)

APPROACHES TO STATE COMPLIANCE WITH INTERNATIONAL TREATIES

The realist perspective in IR theory assumes that states are rational unitary actors that behave mainly to maximize self-interest. Military power and economic power are viewed as the main determinants in international politics. As noted in Chapter 1, realists have not regarded state compliance with international obligations as a particularly interesting issue. For one thing, it is assumed that states generally comply with such obligations. The argument is that: i) states accept treaties only when their governments have concluded that they are in their interest; ii) therefore states generally comply with treaties; and iii) when they don't, sanctions are employed both to punish offenders and to deter others from violating (Jacobson and Weiss, 1995). In the frequently cited words of the famous realist Hans Morgenthau (1948, p. 229), 'The great majority of the rules of international law are generally observed by all nations.'[4] In essence, observed compliance merely reflects one of the following three situations: i) a hegemonic state has forced or induced a less powerful state to comply; ii) the treaty rules merely codify the existing behaviour of the parties; or iii) the treaty resolves a coordination game in which no party has any incentive to violate the rules once a stable equilibrium has been established (Mitchell, 1994a).

In his typology of theories of compliance with international law, Burgstaller (2005) differentiates among realist, institutionalist and normative approaches, and, in turn, divides realist theories into three groups. Offensive realism (Mearsheimer, 2001) views international conflict as the result of the anarchy of the international system, not of human nature *per se*, as classical realists such as Morgenthau assumed. Security is scarce and states try to achieve it by maximizing their relative advantage. They comply with international obligations when this is seen as serving their interests, their relative capabilities and external environment taken into account. Defensive realism, by contrast, argues that international anarchy is more benign. Security is often plentiful rather than scarce, and states can normally afford to be more relaxed in their pursuit of relative advantages than offensive realists assume (see, for instance, Walt, 1998). Accordingly, compliance with international treaties can occur in certain situations, even if this is not, seen in isolation, in the strict military or economic interest of the state in question. Finally, neoclassical realists also maintain that states fundamentally seek to shape their external environment; however, they claim that international anarchy is neither Hobbesian nor benign, but rather murky and difficult to read, and recognize that a state's foreign policy is constrained by both international and domestic demands. It is difficult for states to see whether security is plentiful or scarce, and they must interpret their environment as they move along (Burgstaller, 2005, pp. 98–9). In the famous words of Wendt (1992, p. 391), 'Anarchy is what states make of it.'

Institutionalists claim that, while it might be correct that most states comply with most treaty rules most of the time, this is not because enforcement is applied against non-compliers. From their review of more than 100 treaties, Chayes and Chayes (1991) found that the wording of the treaty almost never stipulates formal punitive sanctions. Membership sanctions, which by contrast *are* often included in treaties, are hardly ever invoked to compel compliance. 'In sum, sanctioning authority is rarely granted by treaty, rarely used when granted, and likely to be ineffective when used' (Chayes and Chayes, 1995, pp. 32–3).[5]

Institutionalists oppose the realist view that non-compliance occurs only (or even primarily) because it is in the interest of a particular state to act in that way. Instead of viewing non-compliance as the result of a deliberate maximization of interest, institutionalists argue that many instances of non-compliance are caused by institutional imperfections in the regime in question, such as ambiguity and indeterminacy of treaty language and limitations on the capacity of parties to carry out their obligations (Chayes and Chayes, 1991, 1995). One famous case is Mitchell's (1994a, 1994b) study of the regime controlling intentional oil pollution at sea. This regime consists of two different sub-regimes, one that prohibits discharge of oil in excess of specified limits, and another that requires installation on board tankers of expensive equipment that reduces pollution. Despite the high costs involved for tanker operators, it proved easier to secure compliance for the equipment sub-regime. Mitchell explains this variance with the different institutional features of the two sub-regimes. Compared to the discharge sub-regime, the equipment sub-regime displayed better transparency, provided more potent and credible sanctions, and reduced implementation costs by building on existing infrastructures.

Further, institutionalists argue that bureaucratic processes often favour compliance over non-compliance. In national bureaucracies, economy counts in decisions to comply with a treaty as a matter of standard operating procedure, rather than weighing the costs and benefits each time an issue of compliance arises. Gathering information and securing inter-agency agreement are high-cost activities and can be performed only in relatively important questions (Chayes and Chayes, 1991). In a similar vein, continuous attempts at persuasion and justification contribute to making compliance the natural choice of behaviour, according to institutionalists: '[T]he fundamental instrument for maintaining compliance with treaties at an acceptable level is an iterative process of discourse among the parties, the treaty organization, and the wider public' (Chayes and Chayes, 1995, p. 25).

In order to increase treaty compliance, institutionalists have recommended transparency in the workings of regimes or treaties, and mechanisms for dispute settlement, as well as technical and financial assistance to states that have practical problems with complying (Chayes and Chayes, 1991, 1995; Mitchell, 1994a, 1994b; Weiss and Jacobson, 1998). Transparency, in the sense that information about the performance of the parties to the treaty is available, may contribute to compliance in various ways. It may reassure a party that others are in compliance and provide the basis for embarrassing and shaming a party that departs from treaty norms. Dispute settlement can ascertain whether a party is in compliance, for instance when others claim it is not. Treaties contain ambiguities, and unforeseen circumstances may appear. According to the institutionalist view, compliance will be improved by an effective process for managing disputes, not only in settling controversies among members, but also in resolving questions of interpretation and in adapting treaty norms to changing circumstances. Finally, in some instances, technical and financial assistance to certain states can be necessary to increase compliance with a specific treaty. In essence, 'Instances of apparent non-compliance are treated as problems to be solved, rather than as wrongs to be punished. In general, the method is verbal, interactive, and consensual' (Chayes and Chayes, 1995, p. 109).

Normative theories on compliance also assume that states generally obey international law, but this is ascribed to the moral and ethical obligation of states, derived from considerations of natural law and justice (Burgstaller, 2005, p. 101). Norms are believed to affect the behaviour of states owing to their internal normative nature; the norm itself has certain qualities based on its origin, content and operation in practice that makes states take it seriously (ibid.). For instance, Franck (1990) in *The Power of Legitimacy among Nations* argues that state compliance with international law mainly follows from the 'compliance pull', or legitimacy, of the specific rule. Legitimacy, in turn, is dependent on the determinacy of the rule (its ability to convey a clear message), its symbolic validation in society (the degree to which cultural signals are used as cues to elicit compliance), its coherence with the principles underlying other rules, and its adherence to a normative hierarchy that springs out of an ultimate system-validating rule. Straddling the normative and instrumentalist approaches are theorists who argue that norms determine compliance, but that this is modified through institutional practice. Koh (1997), for instance, sees a state's propensity to comply as the result of how the state has been influenced by other states, normatively and otherwise, in the transnational discursive legal process. Similarly, liberal theory views norms as one of the building blocks of states' interests, and one of many factors that determine their foreign policy, including their willingness to comply with international law (Burgstaller, 2005, pp. 102, 165ff.).

POST-AGREEMENT BARGAINING

An important institutionalist argument in the compliance debate is that negotiation does not end with the conclusion of a treaty. As we have seen, disputes can be resolved, ambiguities in the treaty text clarified and compliance induced through negotiations *after* the treaty has been concluded. The procedures employed may range from simple bilateral negotiations to formal arbitration, whether specifically provided by the treaty or evolving in response to need. Using the term 'post-agreement negotiation', Spector and Zartman (2003) explore the intersection of negotiation theory and regime theory in order to explain how international regimes evolve through a process of continual negotiation:

> Regimes are born through negotiation processes, and they evolve through postagreement negotiation processes. . . . If regimes are an approach used by international actors to resolve mutually troublesome problems, postagreement negotiation is the process that keeps those regimes vital and alive, renewing and revising them as knowledge, problems, interests, norms, and expectations change. (Ibid., p. 4)

Jönsson and Tallberg (1998) attempt to bridge the gap between the compliance literature and bargaining theory in IR by introducing the concept of 'post-agreement bargaining'. Understanding this as a generic term that refers to all the bargaining processes that follow after the conclusion of a treaty (of which compliance bargaining is a sub-category), they hold that traditional conceptions of both compliance and bargaining must be revised in view of this widespread but largely ignored practice. The literature has seen compliance as either an enforcement problem (realist or neoclassical theory) or a management problem (institutionalist theory), while negotiation theory has been preoccupied with the processes leading up to the signing of an agreement. The literature on compliance focuses on member-state actions in the post-agreement phase, while neglecting dynamic processes like bargaining. Negotiation theory, on the other hand, emphasizes processes, but fails to extend this attention to the post-agreement phase. Admittedly, institutionalism (or the 'management school', the term that the authors use to contrast this approach with the realist or neoclassical 'enforcement school') captures the elements of persuasion and iteration, but:

> as much as aspects of compliance bargaining are caught in the inductive sweep of the management school, these bits and pieces do not resemble anything close to an elaboration of the phenomenon. While touching upon explanations of compliance bargaining (e.g. ambitious treaty texts), the process of compliance bargaining (e.g. persuasion, iteration) and the effects of compliance bargaining (e.g. levels of compliance), the management school does not weld these observations into an elaborate conception of the phenomenon as such. (Jönsson and Tallberg, 1998, p. 375)

The enforcement school, in turn, disregards compliance bargaining owing to its level of abstraction. Compliance requires the threat and use of sanctions to an extent that outweighs the expected benefits of free riding. What matters to the enforcement school is the level of enforcement, not the means by which it is carried out: 'Whereas the management school catches elements of post-agreement bargaining thanks to its wide, inductive sweep of international treaty compliance, the enforcement school misses those same elements owing to its deductive and highly parsimonious approach' (ibid., p. 376).

Jönsson and Tallberg (1998, pp. 378ff.) propose three basic questions that should be asked in studies of post-agreement compliance bargaining: (i) What is the essence of compliance bargaining? (ii) What are the causes of compliance bargaining? (iii) What are the effects of compliance bargaining? They define compliance bargaining as 'a process of bargaining between the signatories to an agreement already concluded, or between the signatories and the international institution governing the agreement, which pertains to the terms and obligations of this agreement' (p. 372). In their introduction to the essence of compliance bargaining (pp. 378–82), they state that they deliberately use the wider term 'bargaining' instead of the more narrow term 'negotiation' because they want to include more non-verbal, indirect or implicit communication than the face-to-face verbal communication normally associated with negotiating. They further argue (pp. 382–4) that there are two main causes of compliance bargaining: established violations, and treaty ambiguity. States may violate their obligations because of lack of will or capacity. The signing of an agreement creates a new bargaining situation, where the underlying structure of cooperative and conflictual elements remains intact but where the specific premises may still change. For instance, depending on the relative bargaining power in the pre-agreement phase, the interests of different states may be unequally reflected in the agreement. In the post-agreement phase, these powers, or states' interests, may change. Ambiguity, in turn, may be unintentional or wilful. Sometimes a 'veil of uncertainty' leaves the text open to different interpretations, necessary at the time of treaty conclusion to get all parties on board. Agreement on specific rights and obligations of states is then postponed to the post-agreement phase.

As the title of Spector and Zartman's (2003) book suggests, post-agreement bargaining is all about 'getting it done'. Compliance bargaining is about 'getting it complied with'. At the theoretical level, post-agreement bargaining theory crosses dividing lines between different schools (more about this below). It basically does not ask 'why it's done', but 'how it's done'. It encourages empirical studies, but also claims theoretical relevance: understanding how it's done may shed light also on the question about why it's done.

SIMILARITIES AND DIFFERENCES IN SUBSTANTIVE CLAIMS AND SCIENTIFIC AMBITION

I started out this chapter by stating that I draw on distinctly different bodies of literature, which only to a limited extent communicate with each other but which nevertheless share some similarities. A recurrent feature of the perspectives reviewed here is their preoccupation with the term 'compliance', with the literature on common-pool resource management as a possible exception. Here compliance is more implicit in the discussion, with coordination more at the forefront. The literature on compliance in fisheries draws both on the general literature on individual compliance and on the research on common-pool resource management. Likewise, post-agreement bargaining theory is a sub-category of the IR study of state compliance with international agreements, although it aims to communicate with bargaining theory as well. Notably, there is little mutual reference between the compliance literature at individual and state levels. Perhaps disciplinary boundaries have prevented such communication: political scientists have focused on state compliance with international law (and on how lower-level bodies deal with decisions made at higher levels, which is not at issue in this book), while the study of individual compliance has largely been left to economists, criminologists, sociologists and anthropologists.

In all these bodies of literature, there is a clear distinction between one 'formal' (rationalist) and one 'enriched' model (see Table 2.1).[6] Theory on individual compliance with the law was long dominated by the assumption that people comply with the law only if that is in their economic interest – what Tyler (2006) calls the instrumental perspective. Since the 1970s, and especially since 1990, this perspective has gradually been supplemented by the normative approach to compliance, which draws on a long tradition in sociology and emphasizes the role of ethics and legitimacy in individuals' decision-making about compliance. Research on compliance in fisheries shares this distinction, but it was only with the emergence of normative theory in the mid-1980s that this research began to gain momentum, especially after the turn of the millennium. The study of common-pool resources was largely Hardinian rationalist work until the emergence in the late 1980s of the self-management and co-management perspectives, which I have earlier grouped together as 'cooperative action theory' (Hønneland, 1999d). Likewise, state compliance with international treaties was couched in realist terms until institutionalist and normative theory entered the scene in the 1990s, leading to a vast upsurge of interest in compliance in IR. Since the late 1990s, there have been scattered con-

Table 2.1   Main research traditions in the study of compliance

Research field	'Formal' model	'Enriched' model
Individual compliance with the law (including compliance in fisheries)	Instrumental perspective	Normative perspective
Common-pool research management	Rationalist approach	Cooperative action theory
State compliance with international law (including post-agreement bargaining theory)	Realist neoclassical theory (enforcement school)	Institutional and normative theory (management school)

tributions to post-agreement bargaining theory, which crosses the divide between the 'formal' and the 'enriched' model. Contributions have mainly been rationalist in nature, but leaning on the institutionalist argument that negotiation does not end with the conclusion of a treaty. Post-agreement bargaining also opens up for more diverse explanations of compliance than prescribed by 'formal' models. Basically, this perspective prescribes empirical studies of 'how it's done' that enable us to better understand theoretically 'why it's done'.

Despite some differences among research fields in the temporal development of 'enriched' models alongside the 'formal' ones, the general trend is similar. The modern classics of the 'formal' approaches to compliance appeared up until the late 1960s: Becker (1968) on the effect of deterrence on individual compliance, Hardin (1968) on the need for state coercion to avoid the tragedy of the commons, Gordon (1954) on fishery economics, Olson (1965) on the logic of collective action, and Morgenthau (1948) on the anarchy of the international system. With the exception of IR, where (neo-) realist approaches have continued to dominate the field (especially in the USA), the 'enriched' models have largely taken over since 1990, crowned by the award of the Nobel Prize in Economics to Elinor Ostrom in 2009. The year 1990 saw the publication of two of the modern classics in the 'enriched' approaches to compliance and common-pool resource management, respectively: Tyler (2006) on the role of ethics and legitimacy on individuals' compliance with the law, and Ostrom's (1990) study of self-management systems.[7] Chayes and Chayes's classic in the management (institutionalist) school on compliance came in 1995, Mitchell's the year before.[8] Especially in the study of compliance in fisheries, there has been a significant increase in contributions since the turn of the millennium.

*Table 2.2   Main assumptions, scientific ambitions and theoretical concepts
of the 'formal' and 'enriched' models to compliance*

	'Formal' model	'Enriched' model
Assumptions	– Unitary, rationally calculating actors driven by self-interest – Social logic that will necessarily ensue	– Actor motivation more mixed than assumed by the 'formalists' – Social dynamics less stylized and predictable than assumed by the 'formalists'
Scientific ambitions	– Explain variation in terms of pre-defined assumptions – Determine the optimal level of coercion	– Test all thinkable explanations empirically (including, but not limited to, those proposed by the 'formalists')
Main theoretical concepts	– Self-interest – Coercion (including sanctions)	– Legitimacy – Norms – Institutional design – Communication

The major difference between the 'formal' and 'enriched' models lies in their scientific ambitions (see Table 2.2), but this has repercussions on their substantive claims as well. The instrumental, rationalist and enforcement approaches to individual compliance, common-pool resource management and state compliance with international law are all stylized models that take for granted certain features of human psychology and social dynamics. Instrumentalists and rationalists assume that humans are guided by self-interest (largely short-term), and that violation of the law or the tragedy of the commons will follow unless deterrence or state coercion is introduced. Their scholarly contributions aim at determining the optimal level of such coercion, rather than subjecting these basic assumptions to empirical analysis. Similarly, realists in IR assume that states are driven by self-interest, and that the relative distribution of military and economic power determines whose interests will win in inter-state relations. Again, these basic assumptions are not up for discussion. Realists also engage in empirical studies, but largely in order to explain world politics in terms of power relations. As we have noted, compliance is not held to be a particularly interesting topic of analysis, simply because states are believed to enter into agreements only if compliance is in their own interest. The general similarity between these 'formal' models, then, is the assumption of unitary, rationally calculating actors driven by self-interest, and of a social logic that will necessarily ensue (a crime being committed, a common-pool resource destroyed, an international treaty concluded and subsequent compliance with it). Similar is also the clear dividing line drawn in the scholarly ambition of studying how self-interest, deterrence and power play out in real-world situations, without questioning the basic assumptions of actors' motivations and the social dynamics to which they lead.

By contrast, the normative perspective on individual compliance, cooperative action theory in the study of common-pool resource management, and institutionalist and normative theory (the management school) in IR all encourage a wider empirical focus. To varying extents, they do not reject the assumptions of the 'formal' models outright,[9] but they want to see them empirically tested. And their investigations often indicate that actor motivations are more mixed than assumed by the 'formalists', and social dynamics less stylized and predictable. If self-interest and sanction are keywords in the 'formal' perspectives, then legitimacy, norms, institution and communication occupy the same position in the 'enriched' models. In the normative perspective on individual compliance, it has been documented that in many situations norms, legitimacy and ethics are more important determinants of compliance than is economic self-interest. Cooperative action theory has demonstrated that tragedy is not automatically the outcome when users share a common-pool resource without state interference – often because users feel obliged to follow the guidelines of a system for which they themselves are responsible. Students of compliance in fisheries emphasize the role of legitimacy, ethics, habit and social pressure when they explain why fishers obey the law. The management school on compliance in IR has shown that states comply with their international obligations even when this is obviously not in their material interest, explaining this as a result of institutional mechanisms, persuasion by other states or iteration (compliance becoming the standard operating procedure, thus spilling over from one situation to other situations).

'Formalists' tend not to take the criticism from scholars who prefer the 'enriched' model as a fundamental attack on their scientific endeavour. They admit that their models are just that – models – and that the world may exhibit more variation than these assume. As noted by Burgstaller (2005, p. 96), realism contains some general assumptions about states' motivations, but does not purport to explain their behaviour in great detail or in all cases. Similarly, Brox (1990, p. 234) asserts, 'It has been demonstrated that the common property theory . . . is a very powerful analytical tool. The controversy about it mainly stems from it being misread, by proponents and opponents alike, as a falsifiable hypothesis, and even as a general "law".' In a way, the two camps fix their attention on different points. 'Formalists' prefer theoretical coherence, whereas scholars propagating the 'enriched' model are more concerned with empirical variation. Post-agreement bargaining theory claims relevance to both camps (although it is clearly influenced by the management school in IR and can be placed in the category of the 'enriched' model owing to its willingness to test the basic assumptions of the 'formal' models). It aims not to help garner support for one of the opposing camps, but to explain how outcomes differ according to circumstances: 'Rather than arguing that the power of international laws, regimes and agreements takes on either high values (liberal idealism) or low values (classical realism), post-agreement bargaining can be introduced to explain variation' (Jönsson and Tallberg, 1998, p. 398). That is also where our focus will be fixed in the following chapters of this book.

NOTES

1. The first paragraph of this section is based on Kuperan and Sutinen (1998).
2. Liberal philosophers like Locke were similarly 'suspicious' of common property, but argued that individuals may collectively agree to respect each other's rights without recourse to external centralized power.
3. Admittedly, Young (1979) has a chapter on compliance in fisheries and uses theoretical approaches similar to those of Sutinen, but his discussion mainly concerns state compliance with an international fisheries agreement, not individual compliance.
4. Often cited is also Henkin (1968, p. 47): 'almost all nations observe almost all principles of international law and almost all of their obligations almost all of the time'.
5. This may not be a valid conclusion one and a half decades later.
6. The terms are inspired by Weber's concept of formal models, characterized by logical consistency (see Weber et al., 1978) and King and Sutinen's (2010) concept of an 'enriched theory of compliance', which opens up for more diverse explanations of human behaviour than formal models usually do.
7. Admittedly, Tyler's influence cannot really compare to Ostrom's.
8. As we have seen, Young (1979) was a pioneer with his theoretical elaboration of various bases of individual compliance, which, in his view, also had international application.
9. Ostrom (1990) points to the positive effects of graduated punishment, Jönsson and Tallberg (1998) to those of a 'sanctioning ladder', and Mitchell (1994a, p. 428) even to those of 'potent and credible sanctions'. We also recall that Gezelius (2007) saw formal enforcement as a necessary condition for social pressure towards compliance to be effective in fisheries.

3. Fisheries management in the Barents Sea

This chapter outlines the empirical setting of our case of post-agreement bargaining, in terms of biology, and legal and institutional arrangements. We start with a brief description of the main fish stocks of the Barents Sea – the economic assets which attract the interest of fishers from many countries, and which the coastal states aim to manage sustainably for future generations. Primary focus is on the three stocks that have been managed jointly by Norway and the Soviet Union/Russia since the 1970s: cod, haddock and capelin. We also take a quick look at the fisheries: Who takes part? How are fishing operations conducted? Then follows an outline of the legal setting of these fisheries. What does the law of the sea say about fisheries jurisdiction, and what are the implications for management of the Barents Sea fisheries? Thereafter I provide overviews of the Norwegian and Russian systems for fisheries management. The chapter ends with a more detailed introduction to the bilateral management regime between the two countries, with emphasis on the working form of the Joint Commission, and the thematic issues that have predominated since it was established in the mid-1970s.[1]

THE FISH RESOURCES

The Barents Sea abounds in fish stocks of a wide range of species. The reason for this abundance is the rich plankton production in these waters, providing food for large stocks of pelagic fish – that is, fish living in the upper layers of the open sea. The pelagic fish stocks of the Barents Sea, first and foremost capelin and herring, are in turn the prey of groundfish such as cod, haddock and saithe. Both pelagic fish and groundfish serve as food for sea birds, marine mammals and people. Cod, capelin and herring are key species in the Barents Sea ecosystem. The cod feed on capelin, herring and smaller cod, while the herring feed on capelin larvae. Periods of growth in the cod and herring stocks, accompanied by a reduced capelin stock, tend to alternate with periods of moderate cod growth when there is little herring, but abundant capelin.

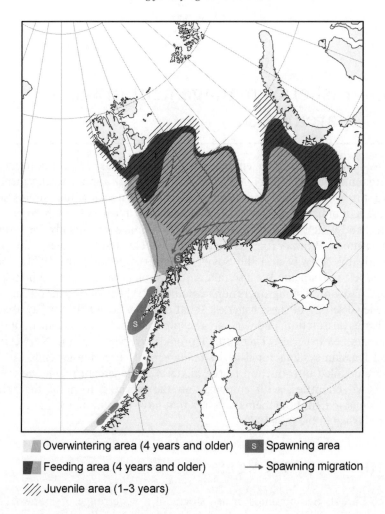

Overwintering area (4 years and older)　　S Spawning area

Feeding area (4 years and older)　　　→ Spawning migration

/// Juvenile area (1–3 years)

Source: Fridtjof Nansen Institute, based on *Havets ressurser og miljø 2009*, Bergen: Institute of Marine Research, 2009.

Figure 3.1　The Barents Sea cod stock

Cod is by far the commercially most important stock in the Barents Sea. North-East Arctic cod (*Gadus morhua*) spawn along the coastline of Norway from the age of seven (see Figure 3.1). After spawning, they return to the Barents Sea. Fry of this species also drift into the northern parts of the Barents Sea. From the age of four, cod prey upon capelin as the latter species moves southwards to its breeding grounds. The cod does not move up to the ice-edge, as the capelin does, but considerable numbers of cod can be found in autumn as far north as west and east of Svalbard. The south-eastern parts of the Barents Sea, from the Kola coast to Novaya Zemlya, are also important breeding grounds for the cod stock. Towards the end of the year, sexually mature cod tend to assemble on the banks off Finnmark before the winter spawning migration. As we will see in the following chapters, the condition of this stock has varied over the years. By and large the situation can be said to be at least satisfactory. As mentioned in Chapter 1, this is the largest cod stock in the world.

The second stock that is seen as shared between Norway and Russia in the Barents Sea is the North-East Arctic haddock (*Melanogrammus aegle-finus*). Haddock is in fact a 'cod fish', that is, it is closely related to cod, and after cod it is the second most important commercial stock in the Barents Sea. However, the haddock stock is far smaller than the cod stock, and its size also tends to fluctuate more widely. Sustaining a stable haddock fishery over time is therefore difficult. Haddock is largely taken as by-catch in the cod fishery (with the exception of fishing with passive gear by small vessels off the Norwegian coast), although haddock may occasionally pre-dominate in the catch. Haddock is found throughout the south-western parts of the Barents Sea (see Figure 3.2). It is found mainly south of the Bear Island – located midway between Svalbard and the Norwegian main-land – but the stock stretches up to the western coast of Svalbard in the north and to the Kola coast in the east. Young haddock are eaten by cod, seal and whales. These predators prefer capelin, so less haddock is preyed upon in periods when capelin is abundant.

The Barents Sea capelin (*Mallotus villosus*) stock was once among the largest fish stocks in the north-east Atlantic. Considerable annual variations in individual growth led to substantial fluctuations in stock size, with implications for the entire ecosystem. The size of the capelin stock remained relatively stable during the 1970s, but fell dramatically in the 1980s. Overfishing reinforced a natural downward trend, bring-ing the stock close to total collapse. Since then, commercial fishing has been allowed only for a few years in succession. Increases in stock size in the early 1990s and around the turn of the millennium were followed by new reductions. Capelin spawn in spring in the near-shore waters of the Barents Sea off Finnmark, the northernmost county of Norway (see Figure 3.3). The fry drift with the current into the northern parts of the Barents Sea, where they mature. Young capelin feed on plankton along the ice-edge from May to October, moving northwards with the melting ice. Towards the end of the year, the mature portion of the stock migrates to the breeding grounds; along the way many become prey to cod that follow the capelin towards the coast. It is largely this annual event that

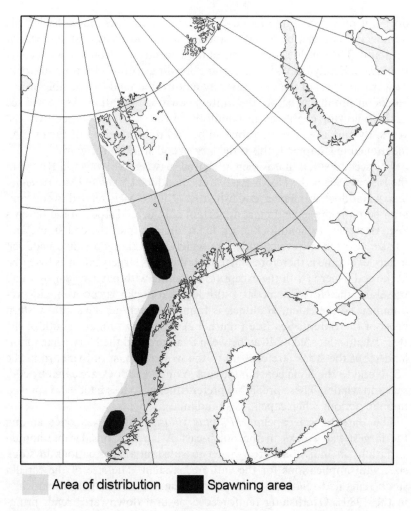

Area of distribution   Spawning area

Source: Fridtjof Nansen Institute, based on *Havets ressurser og miljø 2009*, Bergen:
Institute of Marine Research, 2009.

Figure 3.2   The Barents Sea haddock stock

makes the coast off Finnmark one of the world's richest fishing grounds.
After spawning, most of the capelin die. Fluctuations in stock size thus
depend in part on what proportion of the total stock is sexually mature.
The time it takes for the capelin to reach maturity varies from two to three
years to five or six, depending mainly on sea temperatures.

Atlanto-Scandinavian herring (*Clupea harengus*) was the largest fish

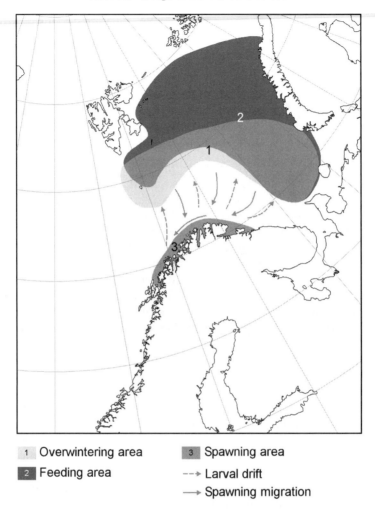

1 Overwintering area   3 Spawning area

2 Feeding area   - - → Larval drift

                                         → Spawning migration

Source: Fridtjof Nansen Institute, based on *Havets ressurser og miljø 2009*, Bergen: Institute of Marine Research, 2009.

Figure 3.3 The Barents Sea capelin stock

stock in European waters until its collapse in the late 1960s. The normal migration pattern from the North Sea to the spawning grounds off the Norwegian coast was interrupted in 1970. Stocks had then become so reduced that the remaining fish could find sufficient food off the Norwegian coast. The old migration pattern resumed only in the mid-1990s. The stock has increased considerably since then, and herring catches in the North Sea have approached those of the 1950s. Some herring fry drift with the Gulf Stream into the Barents Sea each year, although their numbers are highly variable. Each age group spends three years there, feeding on capelin while they drift northwards. Although they contribute to the total herring stock, the young herring of the Barents Sea may be considered a threat to the capelin. However, the herring also relieves the pressure on capelin, since it is itself preyed upon by cod. Herring is not commercially exploited in the Barents Sea, but its presence does affect fishing in the area.

Other important groundfish stocks in the Barents Sea include saithe (*Pollachius virens*; English names also include coalfish and pollack/ pollock), redfish (*Sebastes mentella* and *Sebastes marinus*) and Greenland halibut (*Reinhardtius hippoglossoides*). Traditionally these had been exclusively Norwegian stocks, but changing wandering patterns and new scientific data have led to greater Russian rights to some of these stocks. Russian saithe quotas were increased significantly during the 2000s, as a more plentiful stock began to extend eastwards into Russian waters. Scientific research documented more Greenland halibut in the Russian economic zone than previously assumed, so in 2009 the Joint Norwegian– Russian Fisheries Commission declared this stock the fourth joint stock in the Barents Sea.

THE FISHERY

The Norwegian fishery sector consists of a large number of actors with partly diverging interests. The main groups include the ocean-going fishing fleet, the coastal fishing fleet and the land-based fish-processing industry. The ocean-going fleet consists of a relatively limited number of vessels. The most advanced of them, like the factory trawlers, are mainly registered in the western parts of Norway. The Norwegian coastal fishing fleet consists of a large number of small vessels that fish with conventional gear. Most of these boats are registered in northern Norway. The trawlers' share of the Norwegian cod quota varies with the size of the total allowable catch (TAC) – from 20 per cent when the TAC is low, to 35 per cent when it is high.

After the Russian Revolution of 1917, the north-western fishing industry developed rapidly, concentrated in the Kola Peninsula. Murmansk Trawl Fleet and Murmansk Fish Combine (the latter was to be the largest fish-processing plant in the Soviet Union) were established in the 1920s. This 'northern basin' became the second most important fishing region in the Soviet Union, after the 'far eastern basin'. More recently, a sharp reduction in catches, and subsequently in workers, set in after the economic reforms in Russia that followed the dissolution of the Soviet Union. The vessels of the northern basin, which had until then conducted extensive distant-water fishery off the coasts of Africa and South America, were left with the Barents Sea, closer to home. Catches were now delivered largely in Norway – since the turn of the millennium, in other Western states as well. The north-west Russian fishing fleet has not, however, undergone significant changes in composition. It still consists nearly exclusively of trawlers, although the total number has been reduced from around 400 in late Soviet times to some 250 twenty years later.

Fishing activity in the Barents Sea fluctuates greatly throughout the year and with the species being caught. Among the different nationalities engaged in the fishery, Russian vessels are by far the most numerous. During the most intensive periods, more than 100 Russian trawlers may be fishing for cod in the Barents Sea. The most intensive cod fishery takes place off the coast of Troms and Finnmark – the two northernmost counties in Norway – in late winter and early spring. It continues with lower intensity around Bear Island in late spring and summertime. Shrimp is caught all year round in the waters surrounding Svalbard and close to the mainland, although the Russians have reduced their shrimp fishery in recent years. In years when capelin catches are allowed, considerable Russian fishery for this species is carried out to the south-east of Spitsbergen, the largest island in the Svalbard archipelago.

Fishing activity from Norway and third countries on the open sea is far less intensive. As some 70 to 75 per cent of the Norwegian cod quota is normally allotted to coastal fishers, only a few Norwegian trawlers fish for cod in the Barents Sea, mainly in the Norwegian EEZ. In the waters around Svalbard, shrimp fishery is more prevalent. Unlike the Russians, Norwegian vessels take their capelin quota on the banks close to the Norwegian mainland. Third-country presence in the Barents Sea is also rather limited, at least compared to the fishing activities of Russia. Most intensive in terms of number of vessels is the Spanish summer fishery for cod, usually with a dozen trawlers in the waters around Svalbard. Trawlers from the UK, Germany, France and Portugal normally take up the rest of the EU quota in the Barents Sea, but again this involves significantly fewer vessels than those from Russia.

THE JURISDICTION OF THE BARENTS SEA

Agreement on the principle of 200-mile EEZs was reached at the third UN Conference on the Law of the Sea around 1975. The right and

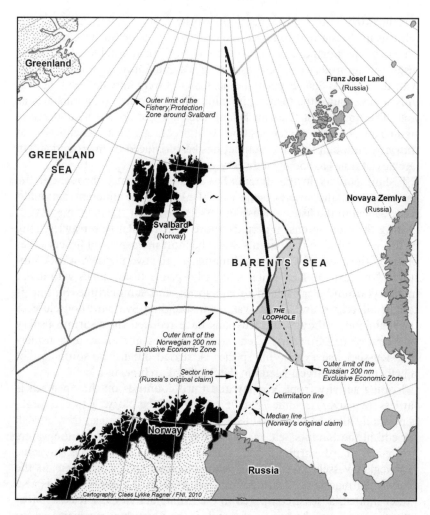

Source: Fridtjof Nansen Institute.

Figure 3.4   Jurisdiction of the Barents Sea

responsibility to manage marine resources within 200 nautical miles of
shore was thus transferred to the coastal states at this time. Both Norway
and the Soviet Union established their EEZs in 1977 (see Figure 3.4).
However, the two states could not agree on the principle for drawing the
delimitation line between their respective zones. The two had been negotiating the delimitation of the Barents Sea continental shelf since the early

1970s, and the division of the EEZs was brought into these discussions. The parties had agreed to use the 1958 Convention of the Continental Shelf as a basis. According to this convention, continental shelves may be divided between states if so agreed. If agreement is not reached, the median line from the mainland border shall normally determine the delimitation line, but special circumstances may warrant adjustments. In the Barents Sea, Norway adhered to the median-line principle, while the Soviet Union claimed the sector-line principle, according to which the line of delimitation would run along the longitude line from the tip of the mainland border to the North Pole. The Soviets generally held out for the sector-line principle, having claimed sector-line limits to Soviet Arctic waters as early as 1926. Moreover, they argued that, in the Barents Sea, special circumstances – notably the size of the Soviet population in the area, and the strategic significance of this region – made it necessary to deviate from the median line. In 1978, a temporary Grey Zone agreement was reached, to avoid unregulated fishing in the disputed area.[2] This agreement required that Norway and the Soviet Union regulate and control their own fishers and third-country fishers licensed by either of them, and abstain from interfering with the activities of the other party's vessels, or vessels licensed by them. The arrangement was explicitly temporary and subject to annual renewal. The Grey Zone functioned well for the purposes of fisheries management,[3] but the prospects of underground hydrocarbon resources in the area pressed the parties to a final delimitation agreement, reached in spring 2010.[4] The agreement is a compromise, with the delimitation line midway between the median line and the sector line.

Another area of contention, with more practical implications for fisheries management in the area, is the Fishery Protection Zone around Svalbard. Norway claims the right to establish an EEZ around the archipelago, but has so far refrained from doing so because the other signatories to the 1920 Svalbard Treaty have signalled that they would not accept such a move. The Svalbard Treaty gave Norway sovereignty over the archipelago, which had till then been a no man's land in the European Arctic. However, the treaty contains several limitations on Norway's right to exercise this jurisdiction. Most importantly, all signatory powers enjoy equal rights to let their citizens extract natural resources on Svalbard. Further, the archipelago is not to be used for military purposes, and there are limitations on Norway's right to impose taxes on residents of Svalbard. The original signatories were Denmark, France, Italy, Japan, the Netherlands, Norway, Sweden, the UK and the USA. The Soviet Union joined in 1935.

The other signatories (apart from Norway) hold that the non-discriminatory code of the Svalbard Treaty must apply also to the ocean area around the archipelago,[5] while Norway refers to the treaty text, which deals only with the land and territorial waters of Svalbard. The waters around Svalbard are important feeding grounds for juvenile cod, and the Protection Zone, determined in 1977, represents a 'middle course' aimed at securing the young fish from unregulated fishing. Most catch restrictions applying to the Norwegian EEZ also apply to the Protection Zone around Svalbard. Separate quotas are not set for this zone. Norwegian and Russian fishers, as well as fishers from third countries, take their catch in whichever zone they prefer as long as they have been licensed for that particular zone – which is usually a formality if the vessel is licensed for fishing in the Barents Sea in general. Each vessel (in Russia: each company) has a fixed quota for the entire Barents Sea. However, fish are bigger closer to the mainland, so Norway prefers that the bulk of the catches be taken in the Norwegian EEZ, in order to reduce potential harm to the stocks. (In addition, enforcement possibilities are naturally better for Norway there than in the other zones of the Barents Sea.)

As follows, the Protection Zone around Svalbard is not recognized by any of the other states that have had quotas in the area since the introduction of the EEZs. To avoid provoking other states, Norway refrained for many years from penalizing violators in the Svalbard Zone. Force was used for the first time in 1993, when Icelandic trawlers and Faroese vessels under flags of convenience – neither with a quota in the Barents Sea – started fishing there. The Norwegian Coast Guard fired warning shots at the ships, which then left the zone. The following year, the first arrest took place in the Svalbard Zone, of an Icelandic vessel fishing without a quota.

Soviet/Russian vessels have been fishing in the Svalbard Zone regularly since its establishment – indeed, they represent the larger part of fishing operations in the area. They do not report their catches in the area to the Norwegian authorities, and Russian captains consistently refuse to sign inspection forms presented by the Norwegian Coast Guard. On the other hand, the Russians do welcome Norwegian inspectors on board, and the same inspection procedures are pursued in the Svalbard Zone as in the Norwegian EEZ. A change in the Norwegian practice of lenient enforcement was first observed in relation to Russian fishers in 1998. The Norwegian Coast Guard decided to arrest a Russian vessel for fishing in an area that had been closed for juvenile-density reasons, but arrest procedures were interrupted before the vessel reached Norwegian harbour, as a result of diplomatic exchanges between the two countries. In 2001, the first arrest of a Russian vessel was carried through in the Protection Zone around Svalbard. Norway claimed the vessel was guilty of serious environmental crime, having violated a number of fishing regulations. Official

Russian reactions were fierce. Russian authorities claimed that Norway had illegally detained a Russian vessel in international waters. Moreover, they accused Norway of breaking a nearly twenty-five-year-old gentlemen's agreement between the two countries, whereby Russia accepted Norwegian monitoring of fishing operations in the Svalbard Zone (including physical inspections of Russian fishing vessels), as long as Norway did not behave as if it had formal sovereignty in the area. The arrest of a Russian vessel was taken as an indication that Norway indeed considered itself as having such formal jurisdiction.[6] The next time Norway attempted to arrest a Russian vessel in the Svalbard Zone was in 2005 – again, for serious violations of fishing regulations, including overfishing. Again the arrest was not carried through, but the reason was different from that in the 1998 incident: the captain of the Russian fishing vessel simply escaped to Russian harbour – taking along the Norwegian inspectors who were still on board. He was later sentenced in Russian court, and official Russian reactions to the Norwegian move were much milder than in 2001. In 2009–10, the Norwegian Coast Guard carried out a handful of arrests of Russian fishing vessels in the Protection Zone around Svalbard, without any formal reaction from the Russian authorities.

Also controversial with respect to fishing rights are the international waters in the north-eastern part of the Barents Sea, the so-called Loophole. Conflicts with Greenland and the EU after Greenlandic and French vessels had started fishery in the Barents Sea Loophole in 1991–92 were solved through diplomatic negotiations. Greenland was allotted a Barents Sea quota, and the EU agreed to include the French Loophole catch in the overall EU quota in the Barents Sea. Icelandic vessels, partly under flags of convenience, started an extensive fishery in the Loophole in 1993. Only in 1999 did the two coastal states reach an agreement with Iceland that put a halt to the Loophole fishery.[7] Like Greenland nearly a decade before, Iceland was given a Barents Sea quota. In return, Norway and Russia acquired fishing quotas in Icelandic waters and received assurances that Iceland would help to combat illegal fishery outside the EEZs in the Barents Sea. Since the turn of the millennium, there has been minimal fishing activity in the Loophole. The area is difficult to access, lying far from shore, and catches are highly variable and usually poor. In effect, fishing in the Loophole is of interest only to vessels without quotas in the Barents Sea. Even the Icelandic fishery in the 1990s had less actual impact on the stock situation than the political interest in the issue would indicate. As noted by Stokke (2010a, p. 60), even at its peak the unregulated Icelandic catch amounted to no more than a third of the *increase* in the total quota from the previous year, and thus posed more of a nuisance than a threat to stock sustainability.

NORWEGIAN FISHERIES MANAGEMENT

Overall responsibility for fisheries management in Norway rests with the Ministry of Fisheries and Coastal Affairs. The main implementing agency is the Directorate of Fisheries, located in Bergen on the western coast. Bergen is also home to the Institute of Marine Research, the main scientific body in Norwegian fisheries management. The institute has been administratively independent of the Directorate of Fisheries since 1988, but it still performs a number of tasks for the Ministry and the Directorate. Notably, it provides management advice based on scientific investigations. The Directorate of Fisheries elaborates proposals for regulatory measures and, following ministerial decisions, implements them by allocating licences, specifying technical regulations, closing certain fishing areas and keeping track of how much individual fishing vessels and foreign states have taken of their quotas. Quotas are distributed by the Ministry of Fisheries following advice from an open Regulatory Meeting, which in 2006 replaced the Regulatory Council, which had been open only to certain actors. The fishing industry has traditionally been heavily involved in Norwegian fisheries management. In the course of the 1990s, environmental organizations were also given a say at these meetings.

The Norwegian Fishers' Association is an important player in Norwegian fisheries management. It represents nearly all Norwegian fishers and is involved in most aspects of the management process. Up until the early 1990s, the association was mainly concerned with negotiations with the Ministry of Fisheries on annual subsidies to the fisheries sector. These subsidies gradually dwindled and vanished altogether when the European Economic Area Agreement entered into force in 1994. Now participation in the process of fisheries management is the primary function of the association. It maintains a continuous dialogue with the authorities about current issues in fisheries management. Since most vessel types are represented in the association, internal conflicts of interest are frequent. On the other side, its near monopoly in representing the interests of Norway's fishers makes it a powerful player in fisheries management. In 1990, some coastal fishers left the association and created the Norwegian Association of Coastal Fishers, which is also consulted by the authorities. The land-based fish-processing industry is represented in the Norwegian Seafood Federation, which includes the country's aquaculture industry.

In the Barents Sea, fishers are subject to control both ashore and at sea (whereas the EU requires control primarily on land). Control measures can be categorized as either passive or active. Passive control refers to official examination of the information that fishers are required to submit about their activities at sea. The authorities 'passively' receive data from the vessels and examine whether there is evidence of lawful behaviour or not. By contrast, active control entails the physical checking by inspectors of this information, at sea or ashore. The passive control of Norwegian fisheries management is exercised by the Directorate of Fisheries, which receives information directly from the fishing vessels (foreign vessels and Norwegian factory trawlers), catch logs (Norwegian vessels exceeding a certain length) and landing data from the fishers' sales organizations (applying to all Norwegian fish-processing plants on shore). On the basis of these data, the Directorate – working together with the sales organizations – keeps continuous track of what remains of the quota of a ship or a foreign nation. Once that quota is exhausted, steps can be taken to halt the fishing. Active quota control is performed by the regional offices of the Directorate of Fisheries when the fish are landed, and by the Coast Guard at sea. The Coast Guard is an integral part of the Norwegian Navy, but performs tasks also for other bodies of governance, such as the Ministry of Fisheries. In practice, fishery inspections are the most important task for the Norwegian Coast Guard. Inspectors board fishing vessels during fishing operations and check the fishing gear and the catch, both on deck (from the latest haul) and in the hold. A main task is to check whether information submitted to the Directorate of Fisheries (through direct reports or through the catch log) corresponds to the actual amounts of fish on board. If, for instance, inspectors find more fish on board than reported to the Directorate, the vessel will be charged with underreporting, which in turn is a reflection of overfishing.

RUSSIAN FISHERIES MANAGEMENT

The Federal Fisheries Agency (*Rosrybolovstvo*) is responsible for fisheries management in the Russian Federation. It was established following the reorganization of the Russian federal bureaucracy in 2004, replacing the Russian State Committee for Fisheries (which in turn had succeeded the Soviet Ministry of Fisheries). Initially placed under the Ministry of Agriculture as a strictly implementing agency, it was again given responsibility for policy formation in 2007, and has been directly subordinate to the government (i.e. not to a ministry) since then. The most important aspect of the 2004 reform was the introduction of three categories of federal bodies of the executive powers, with clear specification of their respective responsibilities, as follows: federal ministries (*ministerstva*) define state policy, federal agencies (*agentstva*) implement this policy and provide services to the population, and federal services (*sluzhby*) perform control and monitoring functions.

From 2004 to 2007, the Ministry of Agriculture was responsible for policy-making in Russian fisheries, while the Ministry's Veterinary Service was responsible for fisheries enforcement (except physical inspections in the Russian EEZ, see below). Since 2007, however, all these functions are again assembled in one body of governance, the Federal Fisheries Agency. This change came after intense lobbying from the Russian fisheries complex. Despite its lower formal status, the Federal Fisheries Agency has wider powers in its specific field than many ministries. A main point of the 2004 reform was to split up political, implementing and controlling functions between different bodies of governance, with the intention of reducing corruption in the Russian bureaucracy, among other things. In fact, the role of the Federal Fisheries Agency is by no means limited to the implementation of government policy. In recent years, the Federal Fisheries Agency has been increasingly active in policy-making and legislative work; it has also regained responsibility for fisheries control, except in the EEZ (see below). Additionally, efforts are under way in the Agency to assume a greater role in policy formation, even lobbying for fisheries to have a specifically dedicated ministry.

The establishment in 2008 of a Governmental Fisheries Commission for development of the fisheries complex is yet another indication of the political will to reform the Russian fisheries sector. According to its statutes, the commission's main role is to ensure efficient cooperation and coordination among and between federal bodies of governance on fisheries-related issues, as well as to consider proposals in fisheries policy, including legislative initiatives. The commission, with its working groups, brings together representatives of interested federal bodies of governance and the fishing industry. It meets at least quarterly and is led by the First Deputy Prime Minister. Since the Federal Fisheries Agency does not have ministerial status, the First Deputy Prime Minister also represents fisheries issues in the government. Until recently, many attempts at improving the regulative framework for the fisheries sector have failed owing to inter-agency tugs-of-war and other conflicts of interest. It is likely that the commission has played an important role in facilitating the adoption and implementation of a considerable number of new regulations previously thwarted by inter-agency differences.

From 2000 to 2003, quota auctions were used as a method of allocating catches. In 2003, the government introduced a fee on quota shares, with quotas allotted for five years in advance, based on the individual ship-owner's proven 'track record' of catch capacity over the last three years (now: five). An inter-ministerial commission under the leadership of the Federal Fisheries Agency now carries out quota distribution of fish stocks that are shared with other states (where TAC is set at the international level, as in the Barents Sea). Amendments to the Federal Fisheries Act in 2007 extended the allocation of quota shares to ten years in order to spur investments in the renewal of Russia's ageing fishing fleet. A second major change, the introduction of mandatory landing in Russia of catches taken in the Russian EEZ, has a three-pronged aim: to secure deliveries of fish to the Russian fish-processing industry, thereby furthering employment objectives; to increase the availability of relatively cheap fish products on the Russian market, furthering nutrition objectives; and to reduce the possibilities for Russian fishers to overfish their quotas, thereby furthering control objectives. Although mandatory landing of catches in Russia does not mean that fishing companies are forced to *sell* their catches to Russian buyers (foreign buyers also operate on the Russian market), a new economic incentive has been introduced to make this option more attractive: a reduction of quota levies to only 10 per cent of the full rate for those who sell their fish on the Russian domestic market.

Traditionally, the Ministry of Fisheries/State Committee for Fisheries has been responsible for all fishery-related issues in Russia, including enforcement of fisheries regulations. In 1997, the President decided to transfer responsibility for enforcement in the Russian EEZ from the State Committee for Fisheries (the predecessor of the Federal Fisheries Agency) to the Federal Border Service (which was incorporated into the Federal Security Service (FSB) in 2003). The Federal Border Service inspects fishing vessels at sea during fishery operations, on the basis of spot checks, or trans-shipment, to see whether the catch log, fishing gear and catch on board comply with the requirements of fishery regulations. The Federal Fisheries Agency and its regional branches have continued to enforce fishery regulations in Russian territorial waters and convention areas – in addition to inland fisheries. The Agency also continues to administer the system for the closing and opening of fishing grounds in cases where excessive numbers of undersized fish are detected in catches.

In 2006–07, a regional branch of the Federal Fisheries Agency was established: the Barents and White Sea Territorial Administration of the Federal Fisheries Agency (the BBTA), which serves as the implementing agency in the northern basin. Quota control in the northern basin is performed by the BBTA, which carries out physical inspections in port and also at sea in Russian territorial waters and outside the Russian EEZ (e.g. in the Barents Sea Loophole and the Fishery Protection Zone around Svalbard). The regional executive authorities in north-western Russia (the governors) established their own fisheries departments in the early 1990s, and had significant influence on quota allocation until quota auctions were introduced in 2000. Since the introduction of the current quota allocation system in 2003, their role has been limited to administering a limited coastal fishery.

Ever since Soviet times, Russia has maintained an extensive system of fisheries research in oceanography, biology of marine organisms, resource assessment, fishing gear and processing technology, among other things. Much of this work was traditionally done outside Soviet/Russian waters to meet the demands of the distant-water fishing fleet. The federal Russian research institute for fisheries is VNIRO, the All-Russian Scientific Institute for Fisheries Research and Oceanography. Regional institutes are found in the various fishery basins, in the northern basin PINRO (Knipovich Scientific Polar Institute for Marine Fisheries Research and Oceanography) in Murmansk, with its affiliate SevPINRO in Arkhangelsk (mainly responsible for marine mammals, inland fisheries and the White Sea fisheries). In the early 1990s, the research institutes became organized as federal state unitary enterprises. This led to a drop in direct financial support, but by way of compensation the institutes received large research quotas – a share of which were fished by commercial fishing companies – to enable research activities at sea, but also for some economic benefit from the sale of catches. In 2007, new regulations made it illegal for the institutes to benefit financially from the research quotas, and public financing of the research institutes was again increased. However, the total outcome seems to have been a net loss for institute budgets of some 20 per cent on average.

There is continuous informal dialogue between Russian fisheries management bodies and the fishing industry, including individual shipowners, associations of shipowners or the processing industry. In the northern basin, the large 'traditional' shipowners like Murmansk Trawl Fleet normally have direct access to government, while the Union of Fishery Enterprises in the North represents some sixty smaller fishing companies (accounting for 30–35 per cent of supplies in the northern basin). A more formal arena for interaction between the Russian fishing industry and the government is provided by the advisory bodies – the so-called fishery councils – found at federal, basin (here: the northern basin) and regional (here: Murmansk county) levels. At the federal level, the Public Fisheries Council was established in 2008 on the basis of the requirement in the Federal Public Chamber Act stipulating that there must be public councils for most federal bodies of governance. Although basin- and regional-level fishery councils have existed since Soviet times, the 2004 Federal Fisheries Act made them mandatory. These councils advise on a wide range of fishery-related issues – including fleet operations; control and surveillance; conservation, recovery and harvesting of aquatic biological resources; distribution of quotas; and other issues of importance to ensure sustainable management of fisheries. The councils are made up of representatives of the fishing industry, federal executive authorities, executive bodies of the Russian federal subjects (the regions), research institutions and some non-governmental organizations, including the indigenous populations of the North, Siberia and the Far East. The current regulations of the Northern Basin Scientific and Fishery Council were issued in 2002. Corresponding regulations for the Murmansk Territorial Fishery Council came in 2005, stating, *inter alia*, that the council is to promote a harmonized fishery policy in the region, and liaise between the fishing industry, fishery authorities, scientific institutions and non-governmental organizations (NGOs).

THE BILATERAL MANAGEMENT REGIME

As explained in Chapter 1, the Joint Norwegian–Soviet (Russian) Fisheries Commission was established by mutual agreement between the two countries in 1975, and met for the first time the year after. The parties had agreed to share equally the two most important groundfish stocks in the Barents Sea, cod and haddock. The Norwegian side had originally proposed 70–30 in Norway's favour, based on the stock's area of dispersion and catch distribution of preceding years. The Soviets argued that in recent years they had deliberately reduced their catches in the Barents Sea, and instead increased their distant-water fishing, precisely to protect the Barents stocks. They found it unreasonable to be punished for this move. The Soviet Union was, after all, a superpower, so the Norwegians agreed to split the quota 50–50.

In the Joint Commission's first few years, fishery relations between Norway and the Soviet Union were dominated by the jurisdictional issues in the area: the establishment of the EEZs, the issue of the Svalbard Zone, and above all negotiations about the delimitation line between Norwegian and Soviet EEZs and the establishment of the temporary Grey Zone. In other words, the work of the Joint Commission in the late 1970s was clearly overshadowed by more high-policy negotiations about ocean governance and jurisdiction between the two countries. The Commission came together two or three times a year, but only for a day or two, often following ministerial meetings about jurisdictional issues. Apart from setting TAC for cod and haddock, it was not yet clear what tasks the Commission would have. At the first session in January 1976, scientific research cooperation topped the agenda. Further, the Soviets proposed collaboration on the artificial breeding of salmon; at that time, aquaculture was less familiar on the other side of the border, so the Norwegian response was lukewarm.

TACs for cod and haddock were set for the first time at the Commission's second session in December 1976. The parties also agreed that Norway could continue fishery with passive gear after its quota share was taken. Following intense negotiations – Norway had demanded 75 per cent, the Soviet Union a 50–50 division – the Commission agreed to split the capelin quota 60–40 in November 1978.

By the early 1980s, fishery relations between Norway and the Soviet Union had taken shape. The parties agreed on which stocks should be managed jointly between them, and on how these stocks should be shared. They had also reached agreement on the exchange of joint stocks and stocks belonging exclusively to one of the parties. Scientific cooperation had become an integral part of the work of the Joint Commission: Norwegian and Soviet fishery scientists had found a new international arena for cooperation in addition to ICES. Other initiatives had been set aside, like cooperation on aquaculture, acclimatization of Pacific Sea salmon in the Barents Sea, and collaboration between the fishing industries of the two countries. Throughout the 1980s, one line of conflict dominated the work of the Joint Commission: Norway wanted to introduce more stringent gear restrictions in the cod and haddock fishery, notably increased minimum mesh size and minimal fish length, while the Soviet Union held out for lower TACs and more restrictions on the Norwegian fishing for spawning cod in the traditional Lofoten fishery. As early as the late 1970s, Norway had proposed that minimum mesh size should be increased from 120 millimetres, which was the NEAFC requirement at the time, to 135 millimetres. The Soviet Union claimed that there was no scientific basis for such a move. Moreover, since fish are generally smaller in the eastern parts of the Barents Sea, this would have had more severe implications for the Soviet than for the Norwegian fishing fleet. In 1982, Norway increased minimum mesh size in the Norwegian EEZ to 135 millimetres – criticized by the Soviets on several occasions in the following years as 'unilateralism' within the context of a joint management regime. In 1980, the Soviet Union agreed to increase minimum mesh size to 125 millimetres, and the parties agreed on minimum fish lengths of 39 centimetres for cod and 35 centimetres for haddock (in 1982 increased to 42 centimetres and 39 centimetres), as well as a ban on more than 15 per cent fish under this length in each catch. Norway regularly overfished its cod quota during the 1980s, but managed to maintain its right to do so as long as the overfishing took place with passive gear. As mentioned in Chapter 1, during the 1980s an exchange scheme was developed whereby Norway was given parts of the Soviet cod and haddock quota shares in return for Soviet quotas on various exclusive Norwegian low-profit stocks. Finally, a working pattern was established during this decade that has been retained up to this day: the Commission meets for one working week in late autumn, after ICES has presented its scientific recommendations for the following year, and occasionally for a shorter session (defined as an integral part of the session that 'commenced' the preceding autumn) in summertime. The sessions are held alternately on the territory of each of the parties. Delegation sizes rose from a mere handful to fifteen–twenty, including representatives of the countries' fishing authorities, ministries of foreign affairs, fishery research institutes and fishing industries.

With the end of the 1980s, a new era for the Joint Commission started. Cod and capelin stocks were in deep crisis, and the lowest TAC in the Commission's history was set for 1990 – 160000 tonnes. (The TAC had varied from 220000 tonnes to around 450000 tonnes during this decade, down from 700000–850000 tonnes in the late 1970s.) Norway gradually abandoned its right to overfish its quota with passive gear. The Norwegian Institute of Marine Research stated that the spawning stock of cod was at its lowest in 120 years, while PINRO in Murmansk recommended a ten-year total ban on fishing cod in the Barents Sea. Whether as a result of the reduced TAC or not, the crisis passed relatively quickly, and the 1990s saw the longest period with sustained high cod TACs, topping at 850000 tonnes in 1997. The good stock situation helped to solve the new dilemma that had arisen in the wake of the introduction of a market economy system in Russia: Russian fishers were no longer interested in the low-price species for which the Soviet Union had traded extra cod and haddock to Norway. With Russian fishers entering the global fish market, it was now Russia that pressed TACs upwards in the Commission, unlike the situation in the 1980s. For the first time, Russian fishers had a real economic incentive to overfish their quotas. As we saw in Chapter 1, Norway uncovered considerable Russian overfishing in 1992–93. Russia's acceptance of the Norwegian accusation, and the ensuing establishment of an enforcement partnership between the two countries, paved the way for the most remarkable development in the Joint Commission during the 1990s: a significant extension in the Commission's area of work. Up till the early 1990s, the Commission had largely been involved in discussions about TAC, quota exchange and the minimal mesh size and fish length. Now the two parties jointly introduced a whole array of new technical regulations, and many existing regulatory measures were coordinated. (This will be discussed at length in Chapter 4.) The arena for most of this work was the Commission's Permanent Committee, which had initially been set up to arrange the enforcement collaboration in early 1993, but became a permanent 'sub-committee' for the Commission later that year. The Permanent Committee consisted of representatives of the countries' fishery authorities, enforcement bodies and fishery science, headed by the Directorate of Fisheries on the Norwegian side and usually by the regional enforcement body in Murmansk on the Russian side. The Permanent Committee met two or three times between each session in the Joint Commission. Contacts between the two parties further increased with the introduction of new electronic means of communication and the establishment of a position as Counsellor for Fisheries at the Norwegian embassy in Moscow in 1997. (Such a position already existed at the Russian embassy in Oslo.) Most notably, direct contact was set up between Norwegian and Russian enforcement bodies, soon extending to daily communication. Annual joint seminars for Norwegian and Russian inspectors were also introduced in 1994. The Commission itself continued to convene twice a year, constituting one session together (now generally held in towns of the fisheries regions of the two countries, such as Murmansk and Tromsø). Delegation size also continued to rise; the Russian delegation to the Commission's session in 1999 consisted of some fifty members. The Russian fishing industry, including several fishing companies in north-western Russia, represented much of this increase, but regional authorities were also represented in the Russian delegation for the first time. A change also took place in the Commission's working form, with more and more of the work being carried out in working groups – for instance, for scientific research, enforcement and fishery statistics. The Commission would meet in plenary session the first day it convened, where all points on the agenda – including the two parties' positions – were briefly presented, while most substantive discussions took place in the working groups throughout the rest of the week. An 'inner circle', consisting of a handful of top officials from each side, dealt with the most important questions – notably the setting of the TAC – and thus represented a kind of high-level working group in its own right. If negotiations were proving particularly hard, the two delegation leaders met one to one, each accompanied only by an interpreter.

The period from the early 1990s to the turn of the millennium marked 'the seven good years' for the Joint Commission. Cod stocks were at an all-time high, and working relations between the parties were exemplary. In general, Norway proposed a range of new management measures, which the Russians then accepted. However, this changed abruptly at the session in Murmansk in November 1999. ICES had recommended a sharp reduction in TAC – from 480 000 tonnes in 1999 to 110 000 tonnes in 2000. Norway was ready to accept this drastic decrease, but the Russian side declared that it had 'not a single fish to give away'. What was worse, the Russians accused Norway of having 'instructed' ICES to recommend such a low quota so as to damage the Russian fishing industry. For the first time in the history of the Commission, negotiations were broken off; not until the night before the Norwegian delegation was to leave Murmansk was the work resumed. The delegation leaders – presumably with a little nudging from the politicians at home – managed to agree on a TAC of 390 000 tonnes for 2000.

Once again, the stock situation improved more quickly than the scientists had forecast, and during the 2000s the TAC for cod was gradually raised to around 600 000 tonnes. A three-year quota for the period 2001–03 and, in particular, new procedural rules for setting the TAC for cod and haddock from 2002 were mechanisms aimed at enabling new compromises between the parties – and new inventions that significantly changed the working form of the Commission. (Both will be further described in Chapter 4.) Previously, the two delegation leaders had spent most of their time during Commission sessions on agreeing on TACs, but now the setting of quotas became more of a technical issue that could be sorted out during the first or second day of the session. Enforcement became the big issue for the Commission in the 2000s, following renewed Norwegian allegations of Russian overfishing. The cooperative atmosphere was far less positive than in 'the glad days of the 1990s', but the result was nevertheless good: by the end of the decade Russian overfishing had been eliminated, according to the Norwegian Directorate of Fisheries. The most remarkable institutional change during this decade was the inclusion of the police, judicial, customs and tax authorities of both countries in the Commission, reflecting the priority given to enforcement issues. In the Russian delegation, the fishing industry and the regional authorities lost most of the influence they had gained during the 1990s. Russian delegations slimmed down again after the turn of the millennium: during the 2000s, some twenty to thirty people normally represented each side at the sessions of the Commission.

There has been a high degree of stability in the personnel represented on the Commission (although new institutions have gradually been given seats on the Commission), especially on the Norwegian side, but also on the Russian side, at least at lower level (delegation leaders being more frequently changed than on the Norwegian side). Many people have represented their respective nations on the Commission for a decade or more. With up to thirty delegation members from each side, as well as a sizeable number of observers and assistants from the country where the meeting is held, plenary sessions of the Commission are large events. When plenary sessions are followed by working group sessions, an established – and for an outside observer quite fascinating – work pattern can be observed: each working group communicates 'upwards' to 'the inner circle' when negotiations have reached deadlock or if the working groups feel they need the approval of their heads of delegation, and 'downwards' to the protocol working group when agreement has been reached on an issue. The protocol group – consisting of several members of each delegation and with its own interpreters – then decides how the decision shall be formulated in the protocol, which is then accepted (or revised) by the delegation leaders. On each working day, each national delegation also holds at least one meeting, where information is exchanged and views on future decisions tested out. On the whole, when the Joint Commission convenes, the venue is more reminiscent of a large international conference than traditional bilateral negotiations.

NOTES

1. The information given in this chapter mainly builds on the author's practical experience from Norwegian and Russian fisheries management; see the section 'Methodological considerations' in Chapter 1. Further information about the Norwegian fishing industry and system for fisheries management can be found in Hersoug (2005) and Hoel (2005). For presentations of Russian fisheries, see Hønneland (2004, 2005) and Jørgensen (2009). Annual overviews of the Barents Sea fish stocks are published by the Norwegian Institute of Marine Research; see, for example, *Havets ressurser og miljø 2009*, Bergen: Institute of Marine Research, 2009. The classic work on the jurisdiction of the Barents Sea is Churchill and Ulfstein (1992). Henriksen and Ulfstein (2011) and Jensen (2011) provide recent updates, covering the 2010 delimitation agreement between Norway and Russia. Other states' policies on Norway's Protection Zone around Svalbard are documented and discussed in a series of articles by Pedersen (2008, 2009a, 2009b, 2011).
2. 'Avtale mellom Norge og Sovjetunionen om en midlertidig praktisk ordning for fisket i et tilstøtende område i Barentshavet med tilhørende protokoll og erklæring', in *Overenskomster med fremmede stater*, Oslo: Ministry of Foreign Affairs, 1978, p. 436.
3. That is to say, there was not much dispute between Norway and the Soviet Union/Russia about how enforcement should be carried out in the Grey Zone. To the extent that there has been Russian overfishing in the Barents Sea, however, it was problematic that the entire area covered by the Grey Zone was inaccessible to the Norwegian Coast Guard (see Chapter 4).
4. *Treaty between Norway and the Russian Federation concerning Maritime Delimitation and Cooperation in the Barents Sea and the Arctic Ocean*, temporarily available at www.regjeringen.no/upload/UD/Vedlegg/Folkerett/avtale_engelsk.pdf.
5. The strongest opposition to the Protection Zone has come from the UK. The USA, Germany and France have formally just reserved their position, which implies that they are still considering their views. Finland declared its support for the Protection Zone in 1976, but has not since repeated it. Canada also expressed its support for the Norwegian position in a bilateral fisheries agreement in 1995, but this agreement has not entered into force. These other Western countries generally accept that the waters surrounding Svalbard are under Norwegian jurisdiction, but they claim that this jurisdiction must be carried out in accordance with the Svalbard Treaty. See Pedersen (2008, 2009a, 2009b, 2011). Russia, on the other hand, formally considers the waters around Svalbard to be high seas. See Vylegzhanin and Zilanov (2007). In practice, however, Russia has accepted Norwegian enforcement of fisheries regulations in the Svalbard Zone. This will be further elaborated in the following chapters.
6. Admittedly, Norway had already carried out the arrest of an Icelandic vessel in the Svalbard Zone in 1994, but Russia supported Norway in the country's battle with

Iceland. This arrest can be viewed as a reaction on behalf of both coastal states in the area towards the extremely serious violation of fishing without a quota.

7. 'Avtale mellom Norge, Island og Russland om visse samarbeidsforhold på fiskeriområ-det', in *Overenskomster med fremmede stater*, Oslo: Ministry of Foreign Affairs, 1999, pp. 838–46.

4. Post-agreement bargaining at state level

When the Soviet Union fell apart in late 1991, the north-west Russian fisheries sector was already undergoing reform, with still more dramatic changes to follow. More and more fishing vessels had begun delivering their catches in neighbouring Norway. Initially an opportunity to earn some hard currency, this practice became a necessity for fishing companies when deliveries in Murmansk became ever more difficult during the 1990s. In just a few months, the Murmansk Fish Combine, once the largest fish-processing factory in the Soviet Union, turned into a 'ghost town' in the harbour of Murmansk. The Norwegian authorities viewed the events with a mixture of glee and horror. On the one hand, the massive Russian catches literally saved numerous fish-processing plants along the Norwegian coast, not yet recovered from the resource crisis of the late 1980s. On the other hand, there were suspicions that Russian enforcement authorities were losing control of how much fish was being taken by their own vessels in the Barents Sea. Soviet quota control had been performed on shore, when the fish was delivered – and the Russian authorities were now deprived of this control opportunity. As we shall see, this was to spur the establishment of enforcement collaboration between Norway and Russia in the Barents Sea fisheries, and subsequent further extension of the bilateral management regime.

This chapter takes us through the rather turbulent post-Cold War period of Norwegian–Russian fisheries relations (turbulent as compared to the immediate past, if not to other fisheries management regimes). It focuses on the biggest issues in fisheries relations between the two countries during these two decades,[1] matters on which Norway has expressly taken a precautionary stance and then attempted to convince Russia to follow suit.[2] We start with the Norwegian suspicions – and subsequent documentation – of Russian overfishing in the early 1990s and the ensuing establishment of enforcement cooperation between Norway and Russia. This new partnership between enforcement bodies from the two countries expanded into broad coordination of technical management measures, and the joint introduction of new ones, during the latter half of the 1990s. As we will see, so far so good: the 1990s were a decade when Norwegian fishery management bodies frequently proposed new management measures – and the Russians routinely accepted them. But then, around the turn of the millennium, Russian rhetoric changed: Norway's initiatives in the Barents Sea were now largely dismissed as being initiated in order to harm Russia; at least, that was what representatives of the Russian fishery management system told the Russian public. Nevertheless, it proved possible to reach compromise between the two countries on several issues, notably the setting of the TAC.

These events are explained in more detail below. I first provide a chronological presentation of each thematic case. This is mainly an account of events as seen from the Commission itself, although I briefly discuss whether its presentation of reality is valid: Was Russian overfishing in the early 1990s actually eliminated? Were TACs brought in line with the precautionary approach during the 2000s? Here I do not provide a full overview of scientific recommendations, established TACs and actual catches each year (see Hønneland, 2006; Stokke, 2010a, forthcoming), but more episodic accounts of central clusters of events during the 1990s and 2000s. My objective is not to document the history of the Joint Commission, but to present the cases in which post-agreement bargaining has been most prevalent.

Then a section on bargaining dynamics follows. Here the focus is not on the outcomes, but on the processes that led to them. What form did Norway's negotiation efforts take? How were the Norwegian initiatives perceived by the Russians? In the final section, I inquire into the effects of the Norwegian initiatives.[3]

As explained in Chapter 1, any discussion of the effects of post-agreement bargaining on Russian behaviour in the Barents Sea fisheries must inevitably touch on 'the big question' of what determines a state's foreign policy. When Russia decides to follow Norwegian initiatives in line with the internationally proclaimed precautionary approach, for example, is that the result of Russia assessing its best interests and behaving accordingly? Does it follow from Norwegian persuasion? If so, which actors were persuaded on the Russian side – the higher political echelons, or sub-groups such as science or enforcement bodies? Or are there institutional features in the Norwegian–Russian management regime that favour responsible management measures? Or general agreement between the two parties to the regime, notwithstanding the desirability of the given outcome from a resource management perspective? One explanation does not necessarily exclude others, and the same empirical data can be used to support different hypotheses. My point of departure (see the section 'Methodological considerations' in Chapter 1) is that it is impossible to 'go into the heads' of people (and far less of states) and report their 'real motives'. The best we can do is to reflect critically on the various possible explanations, based on sound empirical information. Without placing myself in one particular theoretical school, I am definitely open to treating states as more complex entities than rational, unitary actors (again, see Chapter 1).

Finally, while I sporadically raise theoretical questions towards the end of the chapter, the main theory discussion is reserved for Chapter 6.

OVERFISHING IN THE EARLY 1990s

In 1992 – the year the new Russian Federation was implementing drastic economic reform and north-west Russian fishers were delivering ever more fish to Norwegian fish-processing plants – suspicion arose among Norwegian fishery managers that their Russian colleagues were no longer in control of how much fish was being taken up from the Barents Sea by Russian fishers. For one thing, the Russian enforcement authorities had already informed the Norwegian side that their primary source of quota control was the landing of fish. When fish was landed, the authorities checked whether the quantities delivered corresponded to what the vessel had reported over the radio and had written in the catch log (though physical control by inspectors was sparse). How could these enforcement authorities still be in control of the landed quantities now that most Russian-caught fish was delivered outside Russia's borders? Furthermore, the Norwegian Coast Guard revealed – by listening in on Russian catch reports (which took place over the 'open' radio net twice a day) – that Russian fishers were now routinely reporting extremely low catches (normally a mere 0.1 or 0.2 tonnes per haul). Norwegian inspectors could board a vessel shortly after such a radio report and find that the last few hauls in fact contained many times more fish than had been reported, often 1 or 2 tonnes and sometimes even more. The vessel had not violated Norwegian law, since the correct quantity would be inserted in the catch log (at that time, an actual book to be kept on the bridge of the vessel at all times). But the catch log would not be presented for inspection by the Russian enforcement authorities if the vessel was not going to visit a Russian port. This increased Norwegian suspicions that the Russian enforcement authorities had lost control of Russian catches in the Barents Sea. Were they basing their quota calculations on the deflated figures presented by Russian fishers over the radio?

This led the Norwegian fishery authorities to try to calculate the total Russian catch in the Barents Sea that year – not just the quantities taken in Norwegian waters, as they would normally do. They had one important source of information at their disposal: the Russian catch logs. While the Norwegian inspectors might occasionally uncover underreporting of the catch by a Russian vessel (usually a very minor underreporting), the catch logs were generally considered to be a valid source of information: most physical inspections of the holds revealed no discrepancy between fish on board and what had been reported in the catch log. The clue was, during inspection, to go systematically through all information in the catch log from the beginning of the year, including catches taken in the Russian EEZ and the Grey Zone, data that the Norwegians would normally not care about, since enforcement there is the responsibility of the Russian authorities. As most Russian vessels stayed in Norwegian waters for weeks on end, sometimes even months, if the Norwegian Coast Guard could manage to work its way through most of the Russian fleet in the Barents Sea that year it could calculate exactly how much each vessel had fished that year up to the point of inspection.

By the time the Joint Norwegian–Russian Fisheries Commission convened for its annual session in November 1992, the results were clear: the Russians had fished more than 100 000 tonnes of cod above the quota, according to the Norwegian estimate. This estimate was supported by export statistics, which indicated that an amount close to the total Russian cod quota in the Barents Sea had been exported to Norway at the same time as considerable quantities had been exported to other Western countries, and some cod had also been delivered in Murmansk. The Russian cod quota that year was 170 000 tonnes, so the estimated overfishing was significant – nearly 60 per cent.

At the session of the Joint Commission, the heads of the two delegations jointly proposed the appointment of a working group to consider cooperation between the enforcement bodies of the two states. This was reflected as follows in the protocol from the session: 'The Parties agree to appoint a working group consisting of experts in the fields of fisheries regulation, legislation, statistics and control. The working group shall present proposals for concrete cooperative measures by the first quarter of 1993.'[4] No reference was made in the protocol to what was still considered formally to be just an inkling of overfishing. However, after the expert group had submitted its proposals in May 1993, the Joint Commission convened in June for an ad hoc meeting, where the problems related to overfishing and the working group's proposals were the sole items on the agenda. At this meeting, unequivocal reference was made to overfishing in the protocol: 'The Parties noted that a considerable overfishing of cod in all likelihood had taken place in the Barents Sea . . . [and] agreed to ascertain . . . as soon as possible the extent of the overfishing and exchange data on findings.'[5]

The expert group, which had met several times in Norway and Russia during spring 1993, presented a list of eighteen specific proposals within the categories of legislation (2), information (5), control (9) and others (2). Most of them referred to exchange and coordination of information and procedures. In the protocol from the June 1993 session of the Joint Commission, the work of the expert group and its proposals were assessed as follows:

> The Parties expressed great satisfaction with the work of the Norwegian–Russian expert group on cooperation and management of resources. The proposals put forward in the Protocol of 29 May . . . were regarded as very useful. . . . The Parties discussed their further implementation.
>
> The Parties agreed that enforcement authorities in the two countries shall take steps to strengthen enforcement efforts at sea and in connection with landings of catches. There was agreement to establish routines for direct contact between the enforcement authorities of the two countries for exchange of information, concerning practical enforcement routines among other things, so that the Parties in cooperation can make enforcement more effective.
>
> The first meeting of the enforcement authorities of the two countries will take place in Kirkenes [Norway], 15–16 June 1993. As regards the proposal to place observers on each other's vessels, the Parties referred to the planned Coast Guard seminar in Norway, where it will also be possible to discuss further proposals to harmonize enforcement procedures. Further, the Parties agreed to prepare for Norwegian and Russian personnel to accompany vessels used in connection with the closing and opening of fishing grounds.
>
> In accordance with the proposals from the expert group, Norwegian enforcement authorities will require fishery and port permits to be produced on board Russian vessels. In cases in which such documents are not produced, a report will be made to Russian fishery authorities. . . .
>
> The Parties agreed to work for the establishment of a uniform system of conversion factors[6] to be applied by all who fish in the Barents Sea.
>
> The Parties agreed that Norwegian authorities shall convey data on landings of Russian vessels in Norwegian ports to Russian authorities.[7]
>
> The Parties agreed on the necessity to establish direct contact between the official sales organizations so that payment for Russian catches landed in Norway can take place through official sales organizations.
>
> Further, the Parties agreed to exchange upon request copies of weekly catch reports and information about inspections.
>
> The Parties plan to establish a system of data exchange with the help of modern information technology. The practical details will be clarified by experts from the two countries.
>
> The Parties agreed to exchange routinely complete texts of laws and regulations of current interest with a view to an enhanced understanding of each other's fisheries management systems and a possible harmonization of the regulations in certain fields.[8]

Hence, the parties agreed to establish means to enable direct contact between their various enforcement bodies, and exchange law texts, observers and catch data. They also agreed on more ambitious initiatives, such as the elaboration of uniform conversion factors for fish products. Moreover, six months after the Joint Commission had broached the enforcement problems for the first time, the parties agreed on several specific cooperative enforcement measures. For example, the Norwegian enforcement authorities would immediately start forwarding data on Russian landings in Norway to their Russian counterparts, and the Norwegian Coast Guard would start checking Russian fishery permits (giving the captain the right to fish) and port permits (giving the vessel the right to fish).

The first in a series of joint Norwegian–Russian Coast Guard seminars was held at the northern base of the Norwegian Coast Guard at Sortland in September 1993. Two Norwegian cabinet ministers – the Minister of Fisheries and the Minister of Defence – were in attendance, emphasizing the political importance that Norway placed on the joint enforcement deal with the Russians. The protocol of the subsequent session of the Joint Commission in November 1993 refers to the emerging cooperation between the enforcement agencies as follows:

> The Parties expressed satisfaction that in accordance with the Supplement to the Protocol from the twenty-first session of the [Joint] Fisheries Commission, a Coast Guard seminar had been organized at Sortland, where representatives of Norwegian and Russian enforcement authorities discussed improvements in enforcement procedures and future cooperation. The Parties noted that procedures for exchange of information between Norwegian and Russian enforcement authorities have now been established.
>
> The Parties observed that the system for transfer of data on landings by Russian vessels in Norwegian ports functions satisfactorily.
>
> The Parties referred to the fact that the two countries' resource and regulation controls had become more effective through the cooperation of the enforcement authorities of the two countries.
>
> The Parties noted that the installation of modern telecommunications equipment had led to a substantial improvement in the Parties' enforcement activities.[9]

In the years that followed, the Joint Commission was to hail the enforcement cooperation as highly successful, holding it up as an example to be followed by other states.[10] Undoubtedly, the collaboration was successful in the sense that the parties managed to 'do something together'. However, elaborating and implementing regulative measures does not automatically mean solving the problems that spurred their introduction. The 1992 overfishing was related to the new incentive structures facing Russian fishers. Landing fish abroad not only made it *possible* for them to underreport catches: it was only now that it became *profitable* for them to do so. Under the planned economy of the Soviet system, the authorities had had to urge fishers to fulfil their plans, rather than prevent them from overfishing their quotas. Since goods were considerably scarcer than purchasing power, minor rewards in roubles were not sufficient to make overfishing an attractive proposition. The opportunity to land fish abroad suddenly made it considerably more profitable to fish more – even more than the quota allowed – if the difference between real and reported catches could be transformed into money and sent straight into the captain's or fishers' own deposit boxes. Misreporting catches had suddenly become a tempting prospect. The existing Russian enforcement system was proving inadequate, not only because of its lack of information about the catches landed by the vessels, but because of its failure to change the incentives and get the fishers to stay within the law. The old system based on document control in port had probably been adequate in the Soviet period. The few incentives to exploit resources made coercive measures (including physical inspection) unnecessary. Now, with the emergence of such temptations, the shortcomings of the old system became apparent. The core of the enforcement problem did not lie only in the fact that catch data escaped the knowledge of the Russian enforcement authorities; more importantly, there were no social mechanisms in place to persuade fishers to keep within lawful bounds.

The cooperation between the Norwegian and Russian enforcement bodies served the purpose of getting information to the Russian authorities on the activity of Russian fishing vessels in Norwegian waters and ports. Russian fishers could no longer submit incorrect reports about the species and volumes they caught to their own enforcement authorities without being discovered.[11] But does this represent a sufficient step in adapting the enforcement system to its new surroundings? As already indicated, access to information is in itself hardly enough to ensure law-abiding behaviour among fishers. Coercive measures necessarily rely on the ability to enact sanctions in the event of violations. Whether this actually happened in Russia in the 1990s is largely an open question. Indeed, we can question whether fishers feared the sanctions that Russian enforcement bodies had in their arsenal: how threatening was a moderate fine in roubles when the shipowner, captain and crew could expect considerable incomes abroad, hefty at least by Russian standards? In sum, then, the exchange of catch data between Norwegian and Russian enforcement authorities provided one link in the chain of necessary measures to avoid overfishing.

The extent of overfishing after 1992 is in fact uncertain. Official data do not give evidence of any massive overfishing in subsequent years, but they did not do so in 1992 either.[12] Overfishing by Russian vessels that year would not have been discovered had it not been for the efforts of the

Norwegian Coast Guard to calculate the total catch of Russian vessels since the beginning of the year (including their catches in the Russian EEZ and the Grey Zone). Similar calculations were not undertaken for the remainder of the 1990s, as it seems to have been assumed that the problem had been solved. It can even be argued that the Norwegian authorities had an interest in presenting enforcement cooperation with the Russians as a successful joint initiative, and hence would prefer to avoid such calculations in the years ahead. More likely is the interpretation that the Norwegians sincerely believed that the data exchange between Russian and Norwegian enforcement authorities did in fact eliminate overfishing. The good collaborative atmosphere in the Permanent Committee during these years arguably contributed to this (see the section 'Bargaining dynamics' below); the Norwegians did not question the good intentions – or the administrative capabilities – of their Russian collaboration partners. On the other hand, the cod TAC was so high during the last half of the 1990s that overfishing was hardly 'necessary', perhaps not even possible, with the existing catch capacity.

INTRODUCTION AND COORDINATION OF TECHNICAL REGULATIONS DURING THE 1990s

At its session in November 1992, the Joint Commission agreed to appoint a permanent committee which could meet at short notice to discuss management and enforcement issues. The committee was a carry-over from the expert group that had prepared the proposals for enforcement collaboration, and came to be known as the Permanent Norwegian–Russian Committee for Management and Enforcement Cooperation within the Fisheries Sector, or simply the Permanent Committee. The Norwegian delegation to the Committee was composed largely of representatives of the Directorate of Fisheries, but the Coast Guard was also given a seat in connection with the transition from expert group to permanent committee. On the Russian side, the regional enforcement and management bodies in Murmansk dominated, but the Russians also included marine science in their delegation.[13]

The activities of the Permanent Committee from 1994 to the turn of the millennium can be divided into three main categories: i) discussions on current issues related to fisheries management and enforcement practices in the two countries; ii) the administration of exchange of personnel (inspectors and observers) and data; and iii) the execution of more comprehensive tasks assigned to the Committee by the Joint Commission. Most importantly, the meetings of the Permanent Committee enabled the

Norwegian and Russian fishery authorities to discuss issues of current and sometimes urgent interest in greater depth than would have been possible by means of ordinary correspondence. For instance, the Russian authorities frequently requested clarification concerning the interpretation of Norwegian regulations and the enforcement procedures followed by the Norwegian Coast Guard when dealing with Russian fishers. Likewise, current information about the parties' national legislation as well as management set-up and procedures was always on the agenda at the Committee meetings. Finally, the Permanent Committee was the natural place to take up pressing issues, such as the critical situation that arose in 1999 with a massive intermingling of undersized fish in Barents Sea catches.

Exchange of data and personnel was a main proposal to come out of the expert group in 1993 and was one of the Permanent Committee's major concerns in the following years. Procedures and practical arrangements for the exchange of catch and landing data were soon established and functioned without particular problems. Another measure that proved effective was the participation of Russian inspectors as observers on inspections of Russian vessels in Norwegian ports carried out by the Norwegian Directorate of Fisheries. In 1995, the parties agreed to start exchange of inspectors at sea as well. Joint seminars for enforcement officers from the two countries were organized annually from 1993, alternately at the Norwegian Coast Guard base at Sortland, and in Murmansk. These seminars were intended to provide an opportunity for discussions of current issues, and focused mainly on the practical inspection work at sea and in port.

In addition to the administration of the exchange programmes and discussions of pressing issues, the Permanent Committee carried out several larger ventures related to the technical management of fisheries on behalf of the Joint Commission. During the first years, focus was on harmonization of existing Norwegian and Russian management measures. Towards the end of the decade, attention was directed towards the joint introduction of new measures. One major task was the elaboration of common conversion factors for products of fish caught in the Barents Sea. As mentioned above, the introduction of a uniform system of conversion factors was proposed by the expert group in 1993. During the Joint Commission's session in November 1993, the parties agreed 'that it is necessary to elaborate a uniform system of conversion factors, and that the question should be discussed in the Permanent Committee for Management and Enforcement [Cooperation]'.[14] The Permanent Committee appointed a working group to look at the issue. It concluded that some conversion factors were already shared by Norway and Russia; some were slightly divergent; and others differed significantly. In 1994, the Joint Commission ordered the Permanent Committee to 'continue its work to clarify reasons behind the divergent conversion factors, participate in each other's survey cruises and elaborate measuring methods with a view to achieving revised and uniform conversion factors'.[15] In 1995, the Permanent Committee organized two joint Norwegian–Russian survey cruises, and concluded that it should be possible to reach an agreement. In November that year, the Joint Commission requested that the Permanent Committee propose a uniform system of conversion factors and a common method to establish them. Agreement on the method was reached in 1996. The Joint Commission expressed 'great satisfaction with the work carried out [by the Permanent Committee]',[16] and said it would 'adopt the method for use in the future establishment of conversion factors'.[17] Detailed measurement instructions were finalized by the Permanent Committee and adopted for use in the Barents Sea fisheries by the Joint Commission in 1997. Common conversion factors were soon put in place for all the main species found in the Barents Sea.

A second important task for the Permanent Committee was the elaboration of common procedures for the closing and opening of fishing grounds. This is a widely used management measure in both the Norwegian and the Russian parts of the Barents Sea, but the procedures previously employed varied significantly between the two countries. Briefly put: Russian procedures were far more flexible than the Norwegian ones, and there was considerable dissatisfaction among fishers with the way the system worked in Norwegian waters. Decisions on preliminary closure of an area could be made by individual inspectors on the Russian side, whereas in Norway only the Director of Fisheries could make such decisions. Furthermore, the Russians tended to close smaller areas than the Norwegians, and they had more flexible follow-up test hauls after the area was closed. At the meeting of the Permanent Committee in September 1997, the Norwegian delegation requested a briefing about the Russian system for the closing and opening of fishing grounds. During the Joint Commission's subsequent meeting in June 1998, the Commission ordered the Permanent Committee to propose new criteria for decisions on closing and opening fishing grounds. Detailed instructions were drawn up by the Permanent Committee in September. The Joint Commission decided in November the same year to put them into practice on a trial basis. A few minor changes were proposed by the Permanent Committee in May 1999, after which the main framework was in place. This new system brought greater flexibility into the Norwegian procedures for the closing and opening of fishing grounds, based on positive experience with how this worked in Russian waters. It also increased predictability for the fishing fleet; for instance, it became easier for Russian fishers to understand the Norwegian system, now that it was harmonized with the Russian one. Finally, mention should be made of the efforts of the Permanent Committee to standardize the various types of fisheries statistics used in Norway and Russia.

In addition to these major achievements in harmonizing Norwegian and Russian regulation practice, the Permanent Committee played a significant role in paving the way for new management measures, notably the introduction of compulsory selection grids in the cod fishery and satellite communications to track fishing vessels in the Barents Sea. During the Permanent Committee's meeting in September 1994, the parties informed each other about on-going experiments with selection grids, and the Joint Commission that same year noted that tests with grid technology had proved promising. Based on positive results from joint Norwegian–Russian experiments during 1995, the parties of the Joint Commission in November 1995 agreed to introduce compulsory selection grids in cod trawls in specific areas of the Barents Sea as of 1 January 1997. The Permanent Committee coordinated joint experiments and the transfer of test results from the national projects. It produced a set of instructions for the Norwegian–Russian control of selection grids and functioned moreover as a forum to exchange experiences with the use of selection grids.

Tracking vessels by means of satellite surveillance was on the agenda for the first time at the Permanent Committee's September 1996 meeting, at which the parties informed each other about their respective national satellite tracking projects. One year later, the Joint Commission requested the Permanent Committee to assess the possibility of the two countries collaborating on satellite tracking in the Barents Sea. In the Permanent Committee, the parties continued to exchange information on national plans for satellite tracking systems, and in November 1998 the Joint Commission ordered the preparation of a plan to set up a joint Norwegian–Russian satellite tracking initiative in the Barents Sea fisheries. At the Permanent Committee's meeting in May 1999, the parties agreed on an agenda for the first separate meeting with Norwegian and Russian experts on satellite tracking. After a year of intensive meeting activity between these experts and in the Permanent Committee itself, mandatory satellite tracking was introduced for the entire Barents Sea from 2001. This marked the end of nearly a decade of large-scale harmonization of national regulation practices and the joint introduction of new ones between Norway and Russia. The most pressing challenges had arguably been dealt with, but – as we shall see in the section 'Bargaining dynamics' below – the atmosphere between the two countries also changed with the dawn of the new century.

THE PRECAUTIONARY APPROACH AND QUOTA SETTLEMENT AROUND THE TURN OF THE MILLENNIUM

In the latter half of the 1990s, cod quotas in the Barents Sea were at an all-time high, peaking with a TAC of 850 000 tonnes in 1997 – following a gradual increase since the all-time low of 160 000 tonnes in 1990. At the time, marine scientists suspected that their models implied inflated estimates of stock size, and reduced their estimate of total stock size by 200 000 tonnes. In the following two years, the TAC was reduced to 654 000 tonnes and 480 000 tonnes, respectively.

At the same time, the precautionary approach was adopted by both ICES and the Joint Commission. This principle emerged in various regional environmental agreements during the 1980s and became established at the global level in the 1992 Rio Declaration:[18] 'In order to protect the environment, the precautionary approach shall be widely applied by states according to their capabilities. Where there are threats of serious or irreversible damage, lack of full scientific certainty shall not be used as a reason for postponing cost-effective measures to prevent environmental degradation.'[19]

The precautionary approach was subsequently incorporated into international agreements pertaining specifically to fisheries management, notably the 1995 UN Food and Agriculture Organization's (FAO) Code of Conduct for Responsible Fisheries[20] and the 1995 UN Fish Stocks Agreement.[21] The latter declares that states shall apply the precautionary approach, and continues: 'States shall be more cautious when information is uncertain, unreliable or inadequate. The absence of adequate scientific information shall not be used as a reason for postponing or failing to take conservation and management measures.'[22] Hence, the essence of the precautionary approach is that lack of scientific knowledge should not be used as a reason for failing to undertake management measures that could prevent the degradation of the environment or the depletion of common-pool resources.[23] Whereas it was once considered reasonable to take such measures only when it was established with a high degree of certainty that the environment or resource basis would be seriously threatened without such interference, the introduction of the precautionary approach turned the burden of proof upside down: preventive measures should be postponed or omitted only when there was full scientific certainty that they were *not* necessary.

Both the FAO Code of Conduct for Responsible Fisheries and the UN Fish Stocks Agreement prescribe the use of stock-specific reference points as a tool to deal with matters of risk and uncertainty in fisheries management. The latter defines a precautionary reference point as 'an estimated value derived through an agreed scientific procedure, which corresponds to the state of the resource and of the fishery, and which can be used as a guide for fisheries management'.[24] Two types of reference points are singled out: limit reference points and target reference points. While the former set absolute limits to what is considered to be acceptable, the latter imply management goals to aim at. Management strategies are expected to seek to maintain or restore stocks at levels consistent with the agreed-upon target reference point, and to include measures that can be implemented when reference points are approached. It should be a goal for fisheries management systems to ensure that, on average and over time, target reference points are not exceeded. Precautionary reference points are normally set for the size of the spawning stock and for fishing mortality, that is, the share of the stock that perishes for natural reasons or is caught.

In 1996, ICES started work on elaborating reference points for the commercially exploited fish stocks in the north-east Atlantic. Two years later, reference points were set for the North-East Arctic cod stock: the target reference point for the spawning stock was set at 500 000 tonnes, the limit reference point at 112 000 tonnes (which was the lowest observed in the fifty-three-year time series). For fishing mortality, the target reference point was identified as 0.42, the limit reference point as 0.70. This implied that the management of the North-East Arctic cod was held to be in accordance with the precautionary approach only as long as the stock's spawning mass was greater than 500 000 tonnes and the fishing mortality lower than 0.42 on average over an unspecified number of years. Crisis level was reached if the spawning stock fell below 112 000 tonnes or if fishing mortality reached 0.70.

In Norway, the precautionary approach was incorporated in official fishery policy through a White Paper in 1997;[25] in Russia, this principle is still not found in fisheries legislation or policy documents.[26] The Joint Commission never explicitly adopted the precautionary approach as such, but gradually introduced it around the turn of the millennium by adapting its policies to ICES recommendations and technical (if not declaratory) terminology. In the protocol from its 1997 session, the Commission noted:

> The parties agreed on the need to develop further long-term strategies for the management of the joint stocks of the Barents Sea. Until such a strategy is available for cod, the parties agreed that the annual total quota is to be established so that the spawning stock is maintained above 500 000 tonnes at the same time as the fishing mortality in the next years is reduced to less than . . . 0.46.[27]

The same paragraph was used in the protocol from the 1998 session, with the specification that fishing mortality should be reduced to no less than 0.46 by no later than 2001. In the protocol from the 1999 session, the old technical term F_{med} (safe biological limit for fishing mortality) was replaced by F_{pa} (fishing mortality precautionary – i.e. target – reference point) and the aimed-at catch-rate level was reduced from 0.46 to 0.42, that is, brought into accordance with the precautionary recommendation from ICES. In 2000, the Commission requested ICES to reconsider the precautionary reference point for the spawning stock in light of the dynamics of the cod stock over the preceding thirty to forty years. This replicated a unilateral Russian request in a letter to ICES a few months earlier. Although the wording of the letter urged the scientific body to 'reconsider' the reference point, it is clear that both the Russian government and the Joint Commission were in fact calling for a reduction. In 2001, ICES complied with this request and lowered the target reference point for the spawning stock to 460 000 tonnes. At the same time, however, the limit reference point was increased to 220 000 tonnes. Further, the target reference point for fishing mortality was reduced to 0.40, which meant that requirements to precaution became stricter. On the other hand, the limit reference point was raised to 0.74, so here the room for manoeuvre became wider.

Now let us go back to the Joint Commission's establishment of TAC. We have seen there was a significant downward trend in 1998 and 1999, but quotas were still at a reasonable level, as seen from the perspective of the two countries' fishing industries. Then in autumn 1999, ICES sounded the alarm: not only had their models shown inflated estimates of the Barents Sea cod stock; the stock had actually dropped to an alarming level. Spawning stock biomass was believed to be down at 222 000 tonnes, less than half that prescribed by the target reference point and approaching the limit reference point. (This was practically at the new limit reference point established two years later, which by implication is an extremely serious situation for the fish stock.) Seen from the outside, if there ever was a time to make use of the precautionary approach, it was now. ICES's primary TAC recommendation for 2000 – intended to restore the spawning stock to acceptable levels within three years – was 110 000 tonnes, nearly five times less than the 1999 quota. The Joint Commission arrived at 390 000 tonnes, almost four times above scientific recommendations. The following statement is found in the protocol from this session:

> The Norwegian party notes that the level of the cod quota is alarmingly high in consideration of the available stock assessments and the recommendations from ICES. Taking into account the difficult conditions of the population of north-western Russia . . . Norway has nevertheless found it possible to enter into this agreement.[28]

It was clear that precaution had not prevailed in the Joint Commission, and that the Norwegian side was disappointed. Details follow in the section 'Bargaining dynamics' below.

The next year, the Joint Commission made an unexpected move: it established a TAC for three years ahead. This quota was slightly above the quota for 2000 (395 000 tonnes) and was to be applied for 2001, 2002 and 2003, unless the stock situation became worse than expected (in which case the TAC could be reduced) or the precautionary (target) reference points for spawning stock and fishing mortality were reached before the end of 2003 (then the TAC could be increased). The three-year element was obviously intended to provide greater predictability. The fishing industries of Norway and Russia were given better opportunities to plan for the immediate future, and those who feared for the health of the cod stock were given assurances that the TAC would not increase even further unless management objectives had been achieved. Judged by the standards of the precautionary approach, however, much was left to be desired. ICES had recommended a TAC of 263 000 tonnes, and in summer 2001 it declared that fishing mortality of the Barents Sea cod stock could have been as high as 0.9 in 2000.[29] Even the most pessimistic estimates during the Joint Commission's session in November 2000 did not go beyond 0.5.

The new invention announced at the Commission's session in 2002 had far wider implications for the further work of the Joint Commission: a harvest control rule for cod and haddock. The rule consisted of three elements: i) average fishing mortality should be kept below the target reference point for each three-year period; ii) the TAC should not change by more than 10 per cent from one year to the next for cod and 25 per cent for haddock; but iii) exceptions could be made when the spawning stock was below the target reference point. Again we see both biological viability and economic predictability addressed: fishing mortality should be within the precautionary reference point on average for any three-year period, and the fishing industry was secured against large fluctuations in the TAC as long as the spawning stock was above the precautionary reference point. The harvest control rule changed the working form of the Commission. Previously, delegation leaders had spent nearly all their time during sessions negotiating the TACs, and agreement was normally reached only at the very end of a session. Now setting the TAC became more of a technical matter, which could be dealt with sooner rather than later. Perhaps to demonstrate the usefulness of the harvest control rule, the Commission now made a habit of announcing the next year's TAC just a day or two after it convened. As a result, delegation leaders could devote more time and energy to other pressing issues, up till then largely taken care of by the Commission's working groups (see Chapter 3). The harvest control rule was evaluated by ICES in 2005 and found to be in agreement with the precautionary approach. A precondition was that fishing should be brought to a halt – not just reduced – if the spawning stock fell below the limit reference point of 220 000 tonnes.

The cod stock recovered well and the TAC increased gradually during the 2000s, exceeding 600 000 tonnes at the end of the decade. The Joint Commission stuck to its harvest control rule until 2009, when it decided to increase the cod quota by more than 10 per cent and justified the move by referring to the satisfactory state of the stock.[30] The spawning stock was actually above 1 million tonnes at the time.[31] Simultaneously, the Commission added to the harvest control rule that fishing mortality should not be *below* 0.30. Implicitly it should be possible to increase the cod TAC by more than 10 per cent if a quota change within 10 per cent would imply fishing mortality below 0.30. In 2010, ICES evaluated the revision of the harvest control rule and deemed it to be precautionary. The same year the Joint Commission declared that the revised harvest control rule would be used for setting the TAC five years ahead, and then re-evaluated. The cod TAC continued to increase to just above 700 000 tonnes for 2011 – again an increase beyond 10 per cent, but now in compliance with the revised harvest control rule approved by ICES.[32]

OVERFISHING DURING THE 2000s

Disagreement between Norway and Russia around the turn of the millennium about the level of TACs was solved more easily than both parties had probably feared in 1999. With that problem out of the way, however, a new one emerged, at least seen from the Norwegian side. Ever since the enforcement cooperation between the two countries had been established in 1993, there had been little attention to possible discrepancies between the established TAC and actual catches. As noted above, the Joint Commission seemed to assume that overfishing had been eliminated once the Norwegian enforcement authorities started to submit data about Russian landings in Norway to their Russian counterparts. The established routines for enforcement cooperation were codified in a memorandum signed at the Commission's session in 2000,[33] but new challenges were under way. At the meeting of the Permanent Committee in October that year, the parties noted that 'an extensive trans-shipment [of fish] takes place at sea and agreed that this activity is not subject to sufficient control by the parties'.[34] While most north-west Russian vessels had been bringing their fish to Norwegian ports since the early 1990s, the old Soviet practice of trans-shipping fish to transport vessels at sea was resumed. Only now the transport vessels did not go to Murmansk with the fish (as they had in Soviet times), but to other Western countries, like the UK, Denmark, the Netherlands, Portugal and Spain. This was made possible by the gradual upgrading of the north-west Russian fishing fleet to factory trawlers. While fresh fish needs to be landed relatively often – and implicitly to a port close to the fishing ground – frozen products can be kept on board for months, and transported over long distances.

After the Permanent Committee had first pointed out this possible new enforcement challenge in 2000, it was instructed by the Joint Commission to investigate the possibilities for further harmonization of the parties' reporting routines, including exchange of data about their deliveries of fish in third countries. Little came out of this. In 2002, the parties to the Commission declared that they would 'cooperate to produce complete information about landings in third countries'.[35] Further, the protocol said that 'the Norwegian party requested such information from the Russian party' and that 'the Russian party informed that it will continue work to produce data about landings in third countries'.[36] The following year, the parties 'discussed information about unregistered catch of cod in the Barents Sea and the Norwegian Sea'.[37] By 2004, the wording had become tougher: 'With both parties acknowledging that a considerable unregistered catch takes place in the Barents Sea, it is a prioritized goal to use all possible means to reveal and prevent these illegalities.'[38] At the same time, the Russian party noted that Norway, according to official statistics, had a considerable overfishing of cod in preceding years. The Norwegian side explained this by a reform in the regulation system for coastal fisheries, which had hampered its opportunities to halt certain fisheries in time. This overfishing peaked with 21 000 tonnes in 2002 and was subsequently reduced to 10 000 tonnes in 2003 and 4000 tonnes in 2004 (and eliminated by 2005). The Norwegian party, for its part, referred to reports it had submitted to ICES about unregistered catch of cod in the Barents Sea (see below).[39] Following up proposals from the Permanent Committee, the Joint Commission adopted a number of measures that would intensify reporting and control requirements related to trans-shipment at sea: among other things, the obligation for fishing vessels to submit specific reports about all such trans-shipments, the obligation of transport vessels to have satellite tracking devices on board if they receive fish through trans-shipments at sea, and the prohibition against trans-shipment of fish to transport vessels sailing under flags of convenience.[40] At the Commission's session in 2005, the parties agreed to 'continue and ensure the full implementation of measures adopted at the [2004] session',[41] indicating that implementation so far was less than satisfactory. In 2006, the Commission reported that some of the measures agreed upon the previous year had been implemented, others not. Perhaps most ominously, 'the [joint Norwegian–Russian] analysis group that is to put together information at vessel level to reveal possible violations of fisheries regulations has not met during 2006'.[42] And further, 'the sub-committee [on enforcement challenges] under the Permanent Committee . . . has not functioned according to its purpose as there has not been participation from all relevant authorities on the Russian side'.[43] The same formulations appear in the protocol from 2007, with the additional information that the Russian side had failed to appoint a leader to the Permanent Committee's sub-committee on enforcement (which, therefore, did not meet that year either). There was, however, some good news too: 'the parties are pleased to observe indications that the amount of overfishing has been reduced in 2007, among other things as a result of the introduction of the NEAFC port state regime from 1 May 2007'.[44] The same formulation is found in the two following years, while in 2010 overfishing seems to have been brought to a halt:

> The Russian party noted that official fishery statistics show that no overfishing has taken place of Russian quotas of cod and haddock in the Barents Sea and the Norwegian Sea in 2009. . . . The parties noted that the joint Norwegian–Russian effort against overfishing of cod and haddock quotas in the Barents Sea and the Norwegian Sea has brought positive results.[45]

At its session in 2009, the Commission agreed on a joint Norwegian–Russian method for estimating the total catch in the Barents Sea, based on data from satellite tracking and information about transport and landings of fish products.

What is the story behind these protocol formulations? In 2002, the Norwegian Directorate of Fisheries stepped up its efforts to estimate actual Russian catches in the Barents Sea. This unilateral move was the result of what was perceived as lack of interest on the Russian side in the new enforcement challenges (see the section 'Bargaining dynamics' below). An entire new section was built up at the Directorate, recruiting new personnel among experts on economic crime. The section worked systematically on mapping all activity by Russian fishing and transport vessels in the Barents Sea, availing itself of catch reports, satellite tracking data and observations of Russian landings in various third countries.[46] On the basis of this information, ICES estimated unreported catches of North-East Arctic cod as follows: 90 000 tonnes in 2002, 115 000 tonnes in 2003, 117 000 tonnes in 2004, 166 000 tonnes in 2005 and 127 000 tonnes in 2006.[47] These figures implied an annual overfishing in the range of 25–40 per cent of the TAC during the period. In other words, the Russians had been overfishing their national cod quotas by 50 to 80 per cent annually. The Russian fishery authorities, however, did not accept Norwegian assertions that the problem was so severe. In autumn 2006, they admitted not knowing how much fish was actually transferred at sea and delivered to third countries, but estimated Russian overfishing to be around 20000–30000 tonnes annually.[48] According to the Norwegian estimates, overfishing was significantly reduced in the following years: to approximately 41000 tonnes in 2007 and 15000 tonnes in 2008.[49]

While in the protocol from its 2010 session the Joint Commission indicated that overfishing was eradicated through the joint efforts between Norway and Russia, there is general agreement – among experts and in the Norwegian public – that the solution of the problem to a larger extent can be ascribed to the 2007 NEAFC port state regime (Stokke, 2009, 2010b).[50] Under this regime – to which both Norway and Russia are parties – members are not to allow a NEAFC vessel to land or trans-ship frozen fish in its port unless the flag state of the vessel confirms that the vessel has sufficient quota, has reported the catch and is authorized to fish in the area, and that satellite tracking information data correspond with the vessel report. Fifteen per cent of all landings are to be checked physically. NEAFC has also created blacklists of vessels not flying the flag of a state participating in the port state regime that have been observed fishing in the NEAFC Convention Area (including the Barents Sea) without certain evidence that the fish were caught in compliance with NEAFC rules. Such vessels may not land fish to member states or trans-ship fish to vessels from member states.

Hence, around the turn of the millennium, Russian fishers were, to an increasing extent, delivering their catches to transport vessels at sea. The transport vessels largely landed the fish in third countries, in the UK and on the European continent. At least in theory, they could now escape the established enforcement routines between Norway and Russia, as there were no agreements on exchange of landing data with the third countries in question. While such agreements gradually emerged during the 2000s – especially between Norway and various third countries – enforcement of Russian quota regulations ultimately rests with the Russian authorities. The Norwegian authorities can punish a Russian vessel for underreporting catch (i.e. for having more fish on board than indicated in the catch log at the time of inspection), but not for overfishing its annual quota. During the 2000s, the Joint Commission had difficulties agreeing on measures to close the window of opportunity that had opened for fishers to land fish above quota levels; not least, it had difficulties implementing the measures once they were adopted. The enforcement problem was finally solved through multilateral action in NEAFC, with both Norway and Russia on board.

NEW RUSSIAN METHODS FOR ESTIMATING FISH STOCKS

The collaboration between Norwegian and Russian marine scientists is often referred to as the core of the bilateral regime.[51] For one thing, the scientific component of the Norwegian–Russian partnership on fisheries management is the one with the longest history. While collaboration on fisheries regulation started in 1975 and on enforcement in 1993, the first steps towards scientific cooperation had been taken as early as the late nineteenth century (Serebryakov and Solemdal, 2002). Scientists in both countries acquired new large research vessels around the turn of the century and began to exchange information from oceanographic research and observations of fish in their respective national waters and beyond. However, it was only in the 1960s that the Norwegian–Russian/Soviet marine scientific cooperation was formalized (Røttingen et al., 2007). From 1965, joint so-called 0-group investigations were conducted annually, aimed at information about the spawning of all Barents Sea fish stocks. These investigations have continued to this day, probably the longest continuous series of scientific investigations in ICES (ibid.).

According to informants at the Norwegian Institute of Marine Research, interviewed for a research project on knowledge disputes in Russian fisheries science (Aasjord and Hønneland, 2008), Norwegian and Soviet scientists initially had quite divergent perspectives on fisheries research. The Soviet researchers had a more qualitative approach than their Norwegian colleagues, with a stronger focus on ecosystem dynamics. One of our interviewees at the Norwegian Institute of Marine Research described their early cooperation with PINRO as 'a meeting between two traditions; nuanced, but polarized' (ibid., p. 293).[52] The 'Western' approach prevailed, in the sense that PINRO scientists became drawn into the Western-dominated scientific network in ICES: 'We've brought PINRO up to par with us. Now they publish in Western scientific journals. . . . The consequence is that they have become like us, even if that was not our strategy', one Norwegian scientist noted (ibid., p. 293). And another: 'We have, in a way, been their window on the West' (ibid., p. 293). At the same time, the Norwegian scientists expressed ambivalence to their Russian colleagues. According to several interviewees, Norwegian research is scientifically more correct and Norwegian marine scientists are more rational. As one put it: 'We try to maintain integrity. Russian scientists tend to have an agenda. Maybe they're instructed to say something else than what they really mean' (ibid., p. 293).

The evolving scientific cooperation between the Norwegian Institute of Marine Research and PINRO should be viewed in light of more general tendencies in Soviet science. Krementsov (1997, pp. 140ff.) describes the pressure on Soviet scientists to conduct their research in accordance with certain 'patriotic' guidelines and, in particular, the dichotomy between 'patriotic'/Soviet and 'all-human'/'non-patriotic' science, upheld by the Soviet authorities since the late 1940s. After a few years with relatively active contacts between Soviet and Western scientists immediately following the Second World War, the concept of a 'single world science' was abandoned when 'the patriotic campaign' set in around 1947–48. Soviet scientists were no longer to let themselves be influenced by Western bourgeois science,[53] and Western scientists (not to mention intelligence services) should not be allowed to take advantage of the scientific achievements of the Soviet Union. To this end, a ban was put on the publication of Soviet scientific journals in foreign languages and on the translation of abstracts and tables of content into English or other languages, common practice up to then. Patriotism and anti-Westernism became irretrievably linked to Soviet science. Likewise, anybody who failed to display the necessary patriotic values in their research was accused of 'slavishness and servility (*nizkopoklonstvo i rabolepiye*) to the West', a standard epithet at the time (ibid., p. 298). Krementsov (1997, pp. 287ff.) argues that the direction Soviet science took in the late Stalin era was largely a result of the Cold War, and that it remained in force until swept away by the political reforms of the late 1980s. In this context, the evolving East–West cooperation in fisheries science was rather remarkable.

Around the mid-2000s, a schism in Russian fisheries science became evident, through attacks by the federal fisheries research institute, VNIRO, on ICES and the regional institute in the Russian north-west, PINRO. As we saw in Chapter 3, Russia's regional fisheries research institutes in the early 1990s became formally independent of VNIRO, though their scientific work is still reviewed by the federal institute. At the same time, PINRO's relations with the Norwegian Institute of Marine Research expanded, in line with relaxed East–West relations in the European Arctic more widely – and substantial Norwegian funds to support a 'starving' bureaucracy in Russia's north-west. VNIRO had not become part of the international scientific community in ICES to the same extent as PINRO (and had not received financial support from Norway as the regional institute had), and now VNIRO scientists began to question the scientific credibility of the models ICES employed in assessing fish stocks of the Barents Sea. The schism is not mentioned in the protocols from the Joint Commission, but from the early 2000s complaints by VNIRO scientists about ICES models became 'annual performances' at the plenary sessions of the Commission, as expressed by one member of the Norwegian delegation. At first, VNIRO seemed to lack legitimacy in the Russian delegation, at least in its upper echelons, but by the second half of the 2000s Norwegian scientists started to fear that VNIRO's approach would actually prevail on the Russian side.

Exactly what is the essence of the scientific disagreement? In short, VNIRO has considered that the relationship between recruitment to the stock and the size of the spawning stock is given too much weight in ICES models. According to VNIRO, environmental factors (such as natural fluctuations caused by swings in temperature and ocean currents) account for 90 per cent of the conditions for recruitment, and the size of the spawning stock only 10 per cent. In other words, there is no need to be so preoccupied with keeping the spawning stock at a specific level, as its size is of very little importance for how much fish is added to the stock each year. In the preface to a report from a joint Norwegian–Russian scientific workshop in 2006, VNIRO's director stated: '[the] use of completely unreal models which are based on recruitment dependence on abundance of the spawning stock could be treated as *prophesying voodooism* rather than developing scientifically based assessments of the state and dynamics of the fish stocks'.[54]

A central point in VNIRO's criticism of ICES is found in the latter's own figures of the catch pressure on (or fishing mortality of) North-East Arctic cod. Except for a very short period around 1990, fishing mortality has since the 1950s been well above the level that ICES has defined it as necessary to stay below in order to secure long-term viability of the stock, that is, the target reference point. Since the 1970s, fishing mortality has largely been on or above the limit reference point, which according to ICES would represent danger of total collapse of the stock (admittedly only for one in twenty theoretical runs of the entire existing time series for the stock). Well, the stock hasn't collapsed. 'If the reference points and ICES models had been correct, there wouldn't have been any fish in the Barents Sea today', one VNIRO scientist noted in an interview.[55] And he went on: 'The only logical explanation of the divergence between ICES's models and the fact that we still have fish in the Barents Sea is that the estimates are wrong. We underestimate the [cod] stock, and the reason is to be found in the traditional methods.'[56]

VNIRO has presented three models as alternatives to the traditional XSA model used by ICES: the TISVPA model, the GIS model and the 'synoptic' model. Common to these alternative models is that they base estimates on catch reports from the fishing fleet, rather than on data from scientific cruises. Neither catch reports nor cruises cover the entire ocean, so the various models contain different techniques for generalizing from observed amounts of catch to the prevalence of fish in the entire ocean.[57] While the three alternative models are to varying degrees familiar to

ICES – the TISVPA model is, for instance, used as a supplementary tool by several ICES working groups – they all yield significantly higher stock estimates than the XSA model when applied to North-East Arctic cod. In 2006, for example, the XSA model gave estimates of a total cod stock of 1 298 000 tonnes, the TISVPA model of 2 072 000 tonnes, the GIS model of 2 650 000 tonnes and the synoptic model of 2 600 000 tonnes.[58] Hence, two of three alternative models estimated the cod stock in the Barents Sea to be more than twice the size assessed by ICES. This is the crux of the matter for VNIRO: ICES systematically underestimates the cod stock; as a result, the fishing industry loses access to significant amounts of fish. When we met for an interview with the director of VNIRO in December 2007, he received us with a sigh: 'It's horrible what's going on in the Barents Sea at the moment.' Similar statements were common in Norway at the time, referring to the alleged overfishing taking place. However, what the VNIRO director was talking about was '*underfishing*', that is, the failure to utilize the stock's potential.

As mentioned above, the new Russian methods for estimating the Barents Sea cod stock never made it to the protocols of the Joint Commission, but they were included in a draft protocol presented by the Russians at a meeting between the Norwegian Minister of Fisheries and the leader of the Russian Federal Fisheries Agency in March 2006.[59] This does not mean that the alternative methods were never discussed by the Commission: VNIRO routinely promoted the methods at the plenary sessions of the Joint Commission during the 2000s. The big question was to what extent VNIRO represented the official Russian point of view. The leader of the Norwegian delegation to the Commission has repeatedly said that his Russian counterpart has assured him that the new methods will not be applied before they have been thoroughly assessed and accepted by ICES. Norwegian scientists, however, have been more concerned. In May 2006, scientists from the Norwegian Institute of Marine Research told Norwegian media that they felt far from sure that the Russians would not officially promote the new methods at the coming session of the Joint Commission: 'We've been given assurance that they will only be used in connection with symposia, but I feel far from sure.'[60] In our interviews with Norwegian scientists (Aasjord and Hønneland, 2008), similar concerns were expressed: 'The alternative analyses become more and more prevalent. They're like a many-headed ogre. They appear in ever-new variants', one of them said (ibid., p. 303). Another researcher feared that 'someday the Russians will inform Norway that they don't need us any more, but that Norway can still be in the game if we want to' (ibid., p. 303).

In a letter dated 13 October 2006, the Russian Federation requested

ICES to re-evaluate its North-East Arctic cod assessment in view of new information that had become available since ICES had last evaluated the stock a few months earlier.[61] This information included data on Russian trans-shipments at sea – and the synoptic method for estimating the stock size. A group of scientists from Poland, the Netherlands and France was appointed for the task, with designated Norwegian and Russian scientists available to assist. According to *ICES Advice 2006*, there was 'good agreement between the reviewers', and they 'supported the ICES June 2006 advice as they did not find the basis for the "new" stock estimate sufficiently strong to reject the [Arctic Fisheries Working Group] assessment'.[62]

The most outspoken criticism, however, has come not from Norway nor from ICES, but from VNIRO's own former daughter institute in Murmansk, PINRO. In their response to a VNIRO report that presented the synoptic method (Borisov et al., 2006), a group of PINRO researchers (Berenboym et al., 2007) more than hinted that VNIRO had promoted the method for financial rather than scientific reasons: 'The alternative method for estimation of stock size has, even if it was conceived by good intentions, in certain cases been used as an instrument to redistribute research quotas within the framework of existing legislature' (ibid., p. 28). The scientific criticism from PINRO is directed mainly at VNIRO's preoccupation with absolute rather than relative figures:

> One has to remind them that what it is important to know, with respect to rational use of a stock, is not the absolute size of the stock, but how it reacts to the intensity of the fishery. It is not so important whether the absolute size of the stock is 1 million or 10 million tonnes – what is important is how the stock reacts to a certain catch under specific conditions. For example, if an annual catch of 800 000 tonnes from a stock of 1 million tonnes makes it possible to maintain a positive tendency in stock dynamics – without displacing the structure of the stock – then such a catch level can be acceptable. And conversely, if a catch of just 100 000 tonnes from a stock of 100 million tonnes leads to a strong displacement in the stock's structure, then one has to consider this catch level as too high. (Ibid., p. 27)

The PINRO scientists presented their Moscow colleagues as rank amateurs, incompetent in quantitative analysis:

> Until the authors begin to add maximum values of biomass found for different 'synoptic periods' . . . it seems as if one can at least observe a simple logic in their reasoning. . . . But when one comes to the addition of the different maximum biomasses emerging from different time periods, this reminds too much of a pupil's attempts to fit the response to the standard answers in the back of the exercise book. . . . One is amazed at the authors' lack of logic or sophisticated reasoning. (Ibid., pp. 25–6)

BARGAINING DYNAMICS

When the parties to the Joint Commission in late 1992 decided to appoint a working group to consider collaboration between their respective enforcement bodies, the Norwegian side had prepared the ground well. It had produced thorough and convincing documentation of Russian overfishing, based on logbook data from almost the entire north-west Russian fishing fleet. It had also raised the question of a possible overfishing with the Russian side several months prior to the Commission meeting. A member of the Norwegian delegation explained in an interview:

> There was a special meeting [between the parties to the Commission] about overfishing in Moscow in summer 1992. We presented figures on Russian overfishing. [The Russian delegation leader] was aggressive at first, but we emphasized that we wanted to help them tackle the problem of overfishing. So we soon came to agreement.[63]

I was brought into the talks about the emerging enforcement collaboration in early 1993, and was able to attend the meetings in the expert group and subsequently the Permanent Committee, as well as the annual joint Coast Guard seminars, as a participant observer throughout the 1990s (see Chapter 1). From my first meeting in February 1993, I cannot remember ever hearing the Russian delegates question the Norwegian overfishing estimates, or the Norwegian motives for producing them. The same goes for the numerous Norwegian initiatives within technical regulation during this decade. The Russian side generally seemed to accept these suggestions as well founded, and attention was directed towards solving the problems that were identified. Most meetings during the first years were held in Norway, and the Norwegian side put considerable work into familiarizing the Russians with the Norwegian fishing industry and system for fisheries management. Excursions were organized to fishing vessels, fish-processing plants and all the various bodies of governance involved. The Norwegian members of the Permanent Committee invited their Russian colleagues to dinners at their private homes, and personal relations developed. Soon the Russians followed up in Murmansk, with excursions around the city's fishery complex, meetings held at *dachas* in the countryside, and visits to traditional Russian *banyas*. The Norwegian Coast Guard base in the north Norwegian town Sortland – at the time considered to be quite luxurious, popularly known as the Sortland Sheraton – became the primary base for the Norwegian–Russian enforcement cooperation; both parties seemed to feel at home there. The atmosphere in the Permanent Committee was relatively informal; formalities such as official greetings, dinner speeches and elaborate toasts were soon reduced to a minimum (unlike in the Joint

Commission, which maintained a certain level of formality as well as extensive evening programmes). Meetings in the Committee, normally held from Monday to Friday, were above all characterized by hard work. 'Short cuts' were avoided: when the parties discovered that they had opposing views or simply did not understand what the other side meant on an issue, time was set aside for thorough explanation of each party's stance, possible ways ahead and their implications. Often the Committee would resume work after dinner and continue well into the night. The chemistry between the two delegations – and between the two delegation leaders – was good. There was a general feeling of creativity and productivity, spurred not least by the recognition that the Cold War was only a few years behind and that the Committee was breaking new ground also in a wider political perspective.

Around 1998, a small change became evident in the working atmosphere. The Russian delegation leader would now routinely open the meetings with a diatribe: the Norwegians were discriminating against Russian fishers in their inspection activity. Russian fishers were inspected more frequently than fishers from other countries and more severely punished when violations were revealed. The content of these accusations was unexpected for the Norwegian side (although later evidence showed they were not completely unfounded),[64] but most of all it was the form that puzzled the Norwegians. Once the accusation had been delivered, the Russians would return to the 'good working relations' in the Committee, without further mention of the alleged discrimination against Russian fishers. A general assumption emerged in the Norwegian delegation that the Russian delegation leader had been 'instructed by Moscow' to take a tougher stance in the Permanent Committee. Were federal authorities, for instance, suspicious that their own representatives in the Committee were not defending Russian interests well enough? Were these representatives just 'fooling around abroad',[65] becoming a bit too friendly with the Norwegians?

The Norwegian delegation leader to the Joint Commission at the time explained work in the Commission in an interview with me:[66] up until 1997–98, negotiations with the Russians were rather easy. The scientific recommendations allowed for generous TACs, the achievements of the Permanent Committee were considered a feather in the hat also for the Joint Commission, and the Russians were generally flexible at sessions in the Commission:

> In Soviet times, they had a restricted mandate on the Soviet side. From around 1992 to 1997, their delegation leaders decided their mandate themselves, or at least they were allowed to exceed their mandate without prior

consent from Moscow. From 1997, the fishing industry controlled the Russian delegation. . . . In autumn 1997, the scientists cut [the estimates of the total cod stock by] 200000 tonnes in their models. That was the first time I felt that the [fishing] industry had taken control on the Russian side, in 1997 in Petrozavodsk. We ended up 140000 tonnes above the scientific advice. Landings [of fish] were reduced; that was an indication that there was less fish in the ocean and that the scientists were right. But the Russians had difficulties accepting the 200000-tonne cut, and we barely managed to reach agreement. I sat Friday night with [the Russian delegation leader], first in 'the inner circle' [see Chapter 3], then just the two of us with the interpreters, but it was only after we had kicked out the interpreters that we reached agreement. We sat there communicating in simple English; we were at about the same level, language-wise. At last we agreed, and I think I was the one who had to give the most.[67]

My interviewee emphasized the enforcement cooperation and coordination of technical regulation measures as the most important achievements during his time as head of the Norwegian delegation (1989–98). He also underscored that the Norwegians always sought to make the Russians feel ownership of the measures adopted by the Commission on the Norwegian initiative:

The good stock situation and the fact that we could set such high TACs gave us *time* to work with other things than quota issues. In order to make the Russians feel ownership of the measures it was important that things were taken in several rounds: first in the [Permanent Committee] and then in the Commission.[68]

He mentioned the introduction of compulsory selection grids in cod trawls as particularly challenging: 'The selection grid for shrimp was easy. With cod, things were tougher. The Russians understood the Norwegian intention that the selection grid should compensate for the minimum mesh size that we had never succeeded in getting them to accept.[69] But they weren't totally negative, either.'[70] Asked why the Russians accepted the cod selection grid at all, he replied: 'Well, they'd probably "come too far". They had already agreed to so much. They were still in the game, but more hesitant.'[71] Another member of the Norwegian delegation's 'inner circle' at the time also emphasized the *gradual* introduction of selection grids:

The selection grid for shrimp came as early as 1991 or something like that. It was a fisher from Nordmøre who had invented it to rid the trawl of jellyfish. Russian scientists came on board. [A prominent Russian expert on fishing technology] became convinced. It was sort of a mild coercion towards the fishing industry on both sides. . . . The Russians are generally positive, but they are reluctant to reduce the minimum allowable length for shrimp because there's so much small shrimp east in the ocean. This probably opened the way for selection grids in the cod fishery because originally – that is in the shrimp fishery – the grid had hit Norwegian fishers hardest. However, the shrimp selection grid opened up for further exploration of grid technology.[72]

These interviewees admitted that for Norway the selection grids were a way to circumvent Russian reluctance to increasing the minimum allowable mesh size and length of fish or shrimp. They also described the introduction of mandatory grids in cod trawls in 1997 as the final step in a process that had started with the introduction of grids in the shrimp fishery in the early 1990s (allegedly a difficulty primarily for Norwegian fishers[73]) and which gradually bound Russia – if not formally, then in practice – since Russian experts became enthusiastic about grid technology, originally a Norwegian invention. The introduction of grids in the cod fishery was prepared by the Permanent Committee, where the Russian grid experts participated. When the proposal reached the Joint Commission for approval, the process had allegedly come too far for the Russian delegation leader to say no – again: not formally, but in practice. It had become standard operating procedure for the Commission to process – and accept – relatively quickly the proposals that came from the Permanent Committee.

The introduction of grids in cod fishery was soon drawn into emerging Russian complaints that too much had happened too fast in the management of the Barents Sea fisheries. As expressed in a newspaper commentary by a former Soviet delegation leader to the Joint Commission in 1999 (at the time he was working as adviser to the country's fishery authorities at both federal and regional level):

Diplomatic relations with Norway within the fisheries sector have, mildly speaking, been faltering for several years. . . . It all began a couple of years ago with what seemed like a trifle: because of our economic difficulties, we were not able to carry out several joint scientific programmes. Then concessions were made [from our side] in a range of negotiations. Although they might appear insignificant at first glance, they had unfortunate consequences. For instance, we too early agreed to introduce selection grids in the cod trawl according to the Norwegian scheme. . . . Or take the last negotiations on a trilateral agreement between Iceland, Norway and Russia.[74] Yes, this agreement was necessary; no one disagrees on that fact. But it should not have given special rights to the party that is not a co-owner of the Barents Sea resources. Today, our fishers quite rightly raise the question about the Norwegian inspections in the Svalbard area, which are very rigid – sometimes outspokenly biased. For me, who knows the Norwegian fishery control system well, this sometimes seems strange. What worries me is that their inspectors too frequently close fishing grounds with dubious motives – alas, there is too much undersized fish in the catches. . . . There has been a generation shift in our fishery diplomacy. Some have retired because of old age. Others have just 'retired', and, as a result, the logic of our management system has broken down. But that being as it is, does it give the other party the right to take advantage of his partner's mistakes and grab for himself more than is rightfully his?[75]

The Norwegian delegation leader during most of the 1990s (including 1998) referred to this period as 'good times in the Commission'.[76] His successor had a rude awakening in 1999: not only did ICES recommend a fivefold reduction in the TAC for cod; the Russians now declared that they had 'not a single fish to give away'.[77] For the first time in the Commission's history, the session was interrupted. After the plenary session ended on the morning of the session's second day, the head of the Norwegian delegation informed his Russian colleague that he did not see any room for further negotiation. The Norwegian delegation left the negotiation venue at Murmansk Trawl Fleet and spent the next few days at their hotel, discussing the various options available and preparing text for the protocol (on other issues than TAC) in case agreement should be reached at the last minute. That was exactly what happened. Late on Thursday night, the 'protocol group' (see Chapter 3) was rushed to the negotiation venue: the delegation leaders had reached agreement on the TAC, so a protocol had to be produced. This had to be done that same night, as the Russian delegation leader was heading for Moscow the next morning. The Norwegian delegation members outside 'the inner circle' (myself included) were not informed in detail about what had made agreement possible, but it is reasonable to believe that the Norwegian delegation leader had consulted with the political level at home, probably with the Minister of Fisheries. As we saw above, the agreed TAC was closer to the Russian than to the Norwegian preference. The Norwegian Minister of Fisheries had to choose between two evils: non-agreement with Russia, or a TAC far above scientific recommendations.[78]

As we also saw above, the difficult situation at the 1999 session paved the way for new departures as well: first a three-year quota in 2000 and then the harvest control rule in 2002. The head of the Norwegian delegation stressed the connection between what happened in 1999, 2000 and 2002 in an interview with me several years later:

> In 1999, we gave rather much on the Norwegian side in order to get a solution. It wasn't irresponsible, but we had wanted a lower quota. We had had a meeting with the scientists beforehand and took their advice with us to the Commission. The Norwegian goal was already then to achieve long-term management strategies, to get the setting of the TAC 'automatized'. This first led to the three-year quota. The Russians accepted that; we had a good discussion about it. In order to move forward from there we established the Basic Document Working Group, which was given a concrete assignment [in 2001]. Their report [from 2002] gave indications, but no answers. [The Norwegian

Director of Fisheries[79]] and I decided to give it a try. We made the formula and spent several hours talking with [the Russian delegation leader]. He eventually succeeded in getting [the harvest control rule] anchored in the [Russian] group. He gave [a prominent Russian scientist] credit for this.[80]

The Norwegian Director of Fisheries filled in:

It started with a symposium about management strategies that we organized in Bergen in 1998. . . . I had a presentation about stock strategies; a working group was elaborating on the main principles. . . . There had been a system change in Russia. Stability in catches [from the Barents Sea] had now become more important. Previously they had gone in and out of the Barents Sea; now they wanted cod instead of horse mackerel from Far-away-i-stan.[81] So I managed to sneak in something about utilizing the stock on the basis of the [scientific] information available at any time. I presented a draft harvest control rule in a preparatory meeting[82] and spent a lot of time to make [the Russian delegation leader] understand it. After the rule was adopted at the Commission meeting, [the Russian delegation leader] praised one of his own scientists as 'the father of the harvest control rule', during the dinner at night. I interpreted that to the effect that he wanted to cover his back.[83]

The story of how the harvest control rule came about says a good deal about the dynamics both within each national delegation and between them. In Norway, there was growing awareness of the need for long-term sustainable management practices. The context was the emergence of precautionary management guidelines in ICES and a 'tougher climate' in Russia, with the fishing industry acquiring control of the Russian delegation to the Joint Commission. Interacting with the scientific community, a top civil servant worked out a draft harvest control rule that would 'automatize' the TAC setting, paying attention to both biological sustainability and economic viability. He got the Norwegian delegation leader on board and they presented the draft rule to the Russian delegation leader in a smaller context than the Commission itself, at a preparatory meeting. The chemistry was allegedly very good between the two delegation heads at the time,[84] which might have made it easier to reach agreement than if the situation had been less advantageous in that respect. Obviously, the Russian delegation leader felt the need not only to secure legitimacy for the harvest control rule internally in the Russian delegation, but even to present the rule as a Russian invention – in itself arguably a sign that he did not feel convinced about the rule's legitimacy in his own delegation, or in the Russian fishery complex more widely. My Norwegian interviewees indicated that they were by no means insulted that a Russian was given the honour for the harvest control rule; on the contrary, they were pleased to see the rule anchored on the Russian side.

So here we see a line from the Norwegian bureaucracy and scientific community up to the head of the Norwegian delegation, via him to the head of the Russian delegation and down to the Russian scientific community and possibly the rest of the Russian delegation. This does not imply that there is reason to assume any divergence between the scientific communities in Norway and Russia on this issue. As we saw above, scientific convergence is generally high between the leading research institutes in the two countries: the Norwegian Institute of Marine Research and PINRO. The negotiation pattern observed here is probably more a result of the fact that TAC setting had traditionally been the sole responsibility of the delegation leaders. In technical regulation issues during the 1990s, we saw a different pattern: several issues were processed at the technical level in the Permanent Committee – and in ad hoc working groups set up by the Committee, such as on selection grids and satellite tracking – before being presented to the Commission for final approval.

In general, there seems to have been fundamental agreement between the leading scientific communities on both sides, and a tendency to steer towards agreement in the upper levels of the two delegations. The main obstacle seems to have been the Russian fishing industry and actors associated with it, as well as certain lower levels of the Russian bureaucracy. The most difficult TAC negotiations took place in the late 1990s, when the industry had allegedly taken control of the Russian delegation. At the most dramatic session, in 1999, the Russian delegation was headed by a young businessman without experience from the northern basin, who was later accused of economic crime. Here the federal Russian research institute VNIRO might fall into the category of 'actors associated with the Russian fishing industry'; at least, that was PINRO's accusation. As to scientific methods, the established collaboration between PINRO and the Norwegian Institute of Marine Research functioned as a buffer against disagreement between the higher levels of the delegations to the Joint Commission. It would arguably have been more difficult for the head of the Russian delegation – to the extent he was subject to pressure from VNIRO – to stand up against arguments for a new method for estimating the cod stock if the leading Russian research institute on the stocks in question had not been wholeheartedly 'on the Norwegian side'. Russian support might not have been the result of strategic endeavours by the Norwegians to get the Russian scientists 'on their side', but more the consequence of Norwegian efforts to include the Russians in the multilateral scientific community.[85]

The outcome was less favourable for the Norwegians in the overfishing issue during the 2000s. Here the good relations built up between the enforcement bodies of the two countries during the 1990s did not pay off in the form of Russian support for the Norwegian initiative to investigate the consequences of increased trans-shipments at sea. Members of the upper levels in the Norwegian delegation tend to express themselves diplomatically, as the delegation leader did in an interview with me in 2006: 'The Russians acknowledge the overfishing, but they don't present any figures. They're not aggressive or anything like that. They understand that there has been a Russian overfishing; they just don't know how large it's been.'[86] Long-time members of the Norwegian delegation to the Permanent Committee have on several occasions expressed disappointment about Russian lack of initiative in revealing and punishing those who overfish.[87] My own impression from the Commission's working group on enforcement in the mid-2000s was that the parties were not 'on the same wavelength'. The Russians were preoccupied with official statistics: did the figures add up correctly? The Norwegians were more geared towards questioning the validity of these figures.[88] The Norwegians gave top priority to this working group, while the Russians sent low-level civil servants to the meetings. From an interview with a member of the Norwegian delegation to the Permanent Committee:

> There was quite a lot of media stuff on Russian trans-shipment at sea, suspicions of criminality, overfishing. We took up these issues with the Russians, and they said they were worried, too. Indirectly I got the feeling that what they were saying was that they didn't have control. We undertook a common risk evaluation: the Russians had data on trans-shipping, and we had the Coast Guard logbooks – the Coast Guard had registered all trans-shipping for a period. We presented the results in 2000, I think, in the plenary hall [of the northern Coast Guard base], on big spreadsheets. It turned out that only 45 per cent of the trans-shipment observed by the Coast Guard had been reported to the Russian authorities. The atmosphere was, well, not very comfortable . . . the Russians had already been worried, but maybe they didn't realize just what they were in for, with the risk analysis. . . . Quota control is the responsibility of the flag state, but in this case the flag state was evidently not competent at all. Things were floating. . . . About the atmosphere . . . things were straightforward and open from the risk analysis and up until it became clear that the Russians were the only ones with problems here. It was a matter of pride – the great power, and then the little one that came dragging in a lot of muck that no one wanted to know about.[89]

According to my interviewee, the Russian Federal Border Service (see Chapter 3) started to cooperate actively with the Norwegian authorities on the overfishing issue around 2005. A total of fifty-three investigated cases were forwarded from the Norwegian Directorate of Fisheries to the Russian authorities. All twenty-four cases originating in the Russian EEZ, where the Border Service is responsible, were followed up. However, there was no response on the remaining twenty-nine cases. These concerned violations outside the Russian EEZ, where the Russian civilian enforcement authorities (first Murmanrybvod, then the Federal Veterinary Service and then the BBTA; see Chapter 3) were in charge. The latter even withheld data about landings in Norway from the Russian Federal Border Service. My interviewee shared his impressions of the collaboration with the civilian Russian enforcement body as follows:

> [The civilian fisheries authorities] have stayed put [that is, they have not withdrawn from the collaboration with Norway], but it's like sitting in a rowing boat: as long as we're together, we row in the same direction; but when our ways part, they row like mad in the opposite direction. But the atmosphere has been good; even though we've hit them on a sore point, I have never heard one single unpleasant word. As for the civilian track, well, people have been pleasant and all that, but any and every initiative has been stifled, at least in recent years.[90]

He went on to explain that neither the introduction nor the implementation of the joint Norwegian–Russian method for estimating the total catch, adopted in 2009, has been straightforward. Norway and Russia interpret the method's provisions about transparency differently. Section 6.4 in the method requires each party to present 'prepared material' about trans-shipments, transport and deliveries of fish. The Norwegians interpreted this to the effect that there was to be full transparency in all basic documentation. The Russians, however, presented their final analysis and refused the Norwegians access to the basic documentation. The argument was that it contained restricted material from other state agencies, such as the customs authorities. As a consequence, my respondent indicated, one cannot be 100 per cent sure that overfishing has in fact been eliminated:

> Q: Was it difficult to reach agreement on the joint Norwegian–Russian method?
> A: In a way, yes . . . Actually, we wanted to break off before that final meeting; there didn't seem to be much point in going there. But then we were requested to attend. And indeed, there was an agreement . . . but point 6.4 [in the joint method] is the problem. We do not have full access [to basic documentation on the Russian side]. The [official] result is that overfishing no longer exists. We spent time getting the method established, and there is disagreement on how to apply the method. The result is zero overfishing [on paper, at least]. There's been a bit of a tug-of-war and bad atmosphere.[91]

The Norwegians thus had to take the detour through a multilateral organization (NEAFC) in order to get structures created that could eliminate the possibility of Russian fishers delivering fish illegally. In this case, the lower levels of the Russian fisheries bureaucracy allegedly sided more with their own country's fishing industry (or corrupt elements in the fisheries administration) than with similar institutions in Norway. The upper echelons on the Russian side arguably lacked initiative, but enforcement is – unlike the setting of TAC – traditionally handled at lower administrative levels. This situation changed somewhat around 2007–08, when the federal Russian authorities – up to the level of the President[92] – started a campaign to combat illegal fishing activities in Russia (Jørgensen, 2009). This coincided with the elimination of overfishing in the Barents Sea, but was more directed towards the far eastern Russian fishery basin than the country's north-west.

These disagreements and partly abortive initiatives aside, constructive dialogue seems to have dominated sessions in the Joint Commission. As the current head of the Norwegian delegation explained in an interview:

> The tone is basically open and relaxed. Sometimes we might shout at each other, but for the most part there is laughter and pleasant conversation. There is an obvious feeling of community related to the responsibilities we bear. Moreover, seamen are used to helping each other when the need arises, and we see this in the Commission as well. And our neighbourly relations are important – that was especially the case before, when we met more often in north-western Russia. . . . There is an underlying recognition of the closeness of reality.[93]

In the Permanent Committee, collaborative relations soured after the turn of the millennium, especially in recent years:

> We have always been the proactive ones. The initiatives have always come from the Norwegian side. We've taken up the things we felt were wrong. 'Over there', they've always suspected us of having some kind of agenda, wanting to mess things up for the Russian fishers and even for the Russian state.

The main conclusion of my interviewee from the Permanent Committee is, however, that something good has come out of the work and that the atmosphere has been fundamentally good – even though Russian corruption always lurks in the background:

> [The head of the Norwegian delegation to the Permanent Committee from 1993 to 2009] has been very important. She's been like a 'good auntie' for us – and an 'auntie' for the Russians as well. After all, we enjoy being together with the Russians, and that means something. And the Russians enjoy being with us. There have never been any conflicts between us. We [on the Norwegian side] have consciously tried to show understanding for their problems, always. . . . And there are people over there who have suffered, who have been suspended from their positions because they were too cooperative with us. They're the ones who have paid the price for what we have achieved. . . . The fish [that was overfished] we'll never get back, but something positive has come out of that affair, too. . . . They're good people. They're 'like us' – real professionals engaged in the same questions. But it's almost as if there is an ogre lurking in the background all the time.[94]

BARGAINING RESULTS

In Chapter 3, we saw that Jönsson and Tallberg (1998, pp. 378ff.) proposed three basic questions that should be asked in studies of post-agreement compliance bargaining: (i) What is the essence of post-agreement bargaining? (ii) What are the causes of post-agreement bargaining? (iii) What are the effects of post-agreement bargaining? So far in this chapter, I have identified the causes of Norwegian post-agreement bargaining initiatives (like assumed overfishing and tendencies to disregard scientific advice), and I have given some evidence of how the Norwegian side acted – sometimes direct dialogue between delegation leaders, sometimes by encouraging agreement between Norwegian and Russian scientists or technical experts first. But what have been the effects?

Now let us imagine the counterfactual situation that Norway had *not* taken the bargaining initiatives that it did take. Russian overfishing in the early 1990s was suspected and documented by the Norwegians. While we cannot exclude the possibility that Russia might have taken unilateral action to combat overfishing, that would most likely have happened considerably later, if at all. The extensive coordination of technical regulations between Norway and Russia during the 1990s would not have taken place, and selection grids (and perhaps also satellite tracking) would not have been introduced. TACs would in all likelihood have departed significantly more from scientific advice, and the harvest control rule would not have materialized – at least, not in its current form. Overfishing during the 2000s would probably not have been detected, as the Russians were mostly preoccupied with 'paper control'. Even the NEAFC regime might not have emerged, since it, too, was largely a response to the Norwegian estimates of Russian overfishing at the time. Finally, the Russians might have pressed through one of the alternative methods for estimating stocks, such as the synoptic method, if PINRO had not been embedded in the international scientific community.

The overfishing of the early 1990s was probably halted after only a few years (although, as noted above, we cannot be 100 per cent certain about this). Technical regulations *were* coordinated between the two countries, and selection grids and satellite tracking were introduced. TACs were brought closer to ICES scientific advice through Norwegian bargaining, and the harvest control rule was adopted. Russian overfishing was – according to Norwegian estimates – discontinued towards the end of the first decade of the twenty-first century (although there is, in theory at least, insecurity about the current situation now that the joint Norwegian–Russian method has been adopted and the Russians do not give Norway access to basic documentation[95]). And the alternative models for stock estimation never even made it to the protocols of the Joint Commission, although it was more a long-term effect of Norwegian politics that reduced the need for more immediate post-agreement bargaining. While we cannot say for certain that these decisions were 'right', at least they brought the management of the Barents Sea fisheries closer to the standards around which international science – and politics – converges. All this was an effect of Norwegian post-agreement bargaining – understood as 'all those bargaining processes which follow from the conclusion of an agreement' (Jönsson and Tallberg, 1998, p. 372) – in the Commission, in its working groups and in the Permanent Committee and its sub-groups. It came about because Norway did not simply leave implementation of Russia's international obligations (such as conducting precautionary fisheries management) or compliance with adopted measures (such as TACs) to the Russians, but instead viewed compliance bargaining as a continuous aspect of living under the bilateral agreements (see Chapter 2).

This gives rise to several theory-related questions, above all: why did great-power Russia adapt its behaviour to the preferences of small-state Norway? Was it because this specific behaviour was actually in Russia's own interest, as traditional realists would argue? Did Norway help Russia to read the murky environment described by neoclassical realists? Or provide the technical and financial assistance to a state lacking the institutional capacity to comply, as recommended by institutionalists? Did the Norwegian initiatives perhaps have the necessary legitimacy prescribed by normative theory? Or – to return to another institutionalist argument – can Russian compliance be understood as the result of inter-state bureaucratic procedures, established through iteration over time? Had compliance – or agreement between the parties to the Commission – simply become standard operating procedure? We return to these questions in Chapter 6.

NOTES

1. One big issue in the Joint Commission's work since the break-up of the Soviet Union is not discussed here. Around 1997, and for the first time, the Norwegian Institute of Marine Research was refused entry for its research vessels to the Russian economic zone. The Norwegian fishery authorities worked hard during the following years – in the Joint Commission and in day-to-day contact with the Russian authorities – trying to sort out this issue, without much success. A great deal of post-agreement bargaining took place in this case, too, but here the Russian fishery authorities were supportive of the Norwegian requests from the start. The problem lay internally in Russia, with the Northern Fleet (the Federation's largest naval fleet, located on the Kola Peninsula) as the main obstacle. The Russian environmental authorities allegedly also had objections. Presumably, the underlying reasons are to be found in internal power struggles in the Russian bureaucracy. See Hønneland (2004, pp. 58–61) for details.

2. As mentioned in Chapter 1, I do not claim that Norway has necessarily promoted a fishery policy that is 'better' than Russia's in the Barents Sea. However, Norway has more openly declared its support for contemporary international standards for responsible fisheries management, such as the precautionary approach, and has argued that its proposals for specific management measures in the Barents Sea are in accordance with such devices. We return to this point in the present chapter.

3. As mentioned in Chapter 1, the empirical presentation in this chapter builds mainly on protocols from the Joint Commission and its Permanent Committee, and my own (participant) observation in the Norwegian Coast Guard, the Joint Commission and the Permanent Committee. I also build on interviews carried out for my anniversary publication for the Joint Commission to its thirtieth anniversary (Hønneland, 2006), for a research project on knowledge disputes in Russian fisheries science (Aasjord and Hønneland, 2008), and specifically for this book. Sources are indicated only for direct citations and data taken from other sources than protocols and observation. Extracts from protocols of the Joint Commission's sessions are given primarily to provide a flavour of how various problems are spoken of by the Commission itself. The first two sections are shorter versions of Hønneland (2000c). The section on quota settlement builds loosely on Hønneland (2003) and the section on scientific methods on Aasjord and Hønneland (2008).

4. *Protokoll for den 21. sesjon i Den blandete norsk–russiske fiskerikommisjon*, Oslo: Ministry of Fisheries, 1992, Art. 11.2.

5. *Supplement til protokoll for den 21. sesjon i Den blandete norsk–russiske fiskerikommisjon*, Oslo: Ministry of Fisheries, 1993, Art. 3.

6. Factors used to calculate the weight of round fish from the weight of fish products.

7. This measure had been proposed by the Joint Commission one year earlier, but does not seem to have been put in place before the expert group ended its inquiry in the spring 1993; *Supplement til protokoll for den 20. sesjon i Den blandete norsk–russiske fiskerikommisjon*, Oslo: Ministry of Fisheries, 1992, Art. 5.

8. *Supplement til protokoll for den 21. sesjon i Den blandete norsk–russiske fiskerikommisjon*, Oslo: Ministry of Fisheries, 1993, Art. 5.

9. *Protokoll for den 22. sesjon i Den blandete norsk–russiske fiskerikommisjon*, Oslo: Ministry of Fisheries, 1993, Art. 11.2.

10. See, for instance, the protocols of the Joint Commission's sessions in 1995 and 1996: 'The Parties observed that the way in which enforcement cooperation between Norway and Russia is organized is an example for other states to follow.' *Protokoll for den 24. sesjon i Den blandete norsk–russiske fiskerikommisjon*, Oslo: Ministry of Fisheries,1995, Art. 11.2; *Protokoll for den 25. sesjon i Den blandete norsk–russiske fiskerikommisjon*, Oslo: Ministry of Fisheries, 1996, Art. 11.2.

11. This was the Norwegian assumption at the time. As we will see below, there is reason to believe that the Russians never used the information submitted by the Norwegians in their own quota control.

12. Catch data from ICES – considered to be the most reliable source of information about fisheries in the north-east Atlantic – were based on Russia's official reports, not the calculations from the Norwegian Coast Guard. However, ICES later estimated an overfishing of 130 000 tonnes in 1992, 50 000 tonnes in 1991 and 1993, and 25 000 tonnes in 1990 and 1994.

13. The Norwegians invited marine scientists to participate on an ad hoc basis, but not as permanent members of the Committee.

14. *Protokoll for den 22. sesjon i Den blandete norsk–russiske fiskerikommisjon*, Oslo: Ministry of Fisheries, 1993, Art. 11.3.

15. *Protokoll fra den 23. sesjon i Den blandete norsk–russiske fiskerikommisjon*, Oslo: Ministry of Fisheries, 1994, Art. 11.3.

16. *Protokoll for den 25. sesjon i Den blandete norsk–russiske fiskerikommisjon*, Oslo: Ministry of Fisheries, 1996, Art. 11.3.

17. Ibid.

18. It was also included in new global environmental treaties adopted in Rio, such as the Biodiversity Convention and the Climate Convention; see Andresen et al. (2012).
19. 'Rio Declaration on Environment and Development', Rio de Janeiro, 16 June 1992, *International Legal Materials* 31: 874, Art. 15.
20. *Code of Conduct for Responsible Fisheries*, signed in Rome, 28 September 1995, Rome: UN Food and Agriculture Organization.
21. 'Agreement for the Implementation of the Provisions of the United Nations Convention on the Law of the Sea of 10 December 1982 Relating to the Conservation and Management of Straddling Fish Stocks and Highly Migratory Fish Stocks', New York, 4 August 1995, *International Legal Materials* 1547–80.
22. Ibid., Art. 6.2.
23. There was initially considerable scepticism about the precautionary principle in fisheries circles. As noted by Garcia (1994, p. 100): 'The wide adoption of the principle could change drastically the state of affairs in marine living resources conservation and could offer an opportunity to improve fisheries management and ensure sustainable fisheries development. Its careless generalization to fisheries could, however, lead to economic and social chaos in the fishing industry.' As a consequence, the FAO came to advocate the precautionary *approach* for fisheries management as an alternative to the precautionary *principle*, as known in international environmental law. The approach is considered to be a less stringent variant of the principle, entailing a more flexible application of precautionary measures. Hewison (1996, pp. 320ff.) claims that the precautionary approach to fisheries management should involve, among other things, a minimum estimate of stock biomass, a scientifically based and tested management system, a decision-making procedure characterized by transparency and stakeholder (including environmental NGOs) participation, as well as mechanisms to secure compliance among fishers. Garcia (1994, p. 120) names the following steps: improving participation of 'non-fishery users', improving decision-making procedures, and strengthening monitoring, control and surveillance, as well as raising penalties to deterrent levels.
24. 'Agreement for the Implementation of the Provisions of the United Nations Convention on the Law of the Sea of 10 December 1982 Relating to the Conservation and Management of Straddling Fish Stocks and Highly Migratory Fish Stocks', New York, 4 August 1995, *International Legal Materials*, Annex II, Para. I.
25. *St.meld. nr. 51 (1997–98) Perspektiver på norsk fiskerinæring*, Oslo: Stortinget, 1997.
26. See Hønneland (2004, 2005) and Jørgensen (2009) for overviews of Russian fisheries legislation. The 2004 Federal Fisheries Act defines the concept of 'protection and rational use' of aquatic biological resources as the main goal of Russian fisheries management. This concept was widespread in Soviet legislation for the protection of the environment and exploitation of natural resources, and has remained so in the Russian Federation. 'Rational use' might often be given the upper hand over 'protection', but the concept bears some resemblance to the internationally recognized ideal of sustainability, in so far as the emphasis is on long-term and sustained use of the resource, supported by science for socio-economic purposes. The precautionary approach is not found in Russian fisheries legislation and does not seem to have any high standing in Russian fisheries circles, where it is perceived largely as a Western invention. Some even claim it was invented by the West to harm Russia. In our interview with the VNIRO director in Moscow in December 2007, he argued that the 1995 UN Fish Stocks Agreement, which introduced the precautionary approach in international fisheries law, was 'written by Greenpeace with money from the CIA', with the objective of forcing protective measures on Russia while the West built up aquaculture to take over the global fish market.
27. *Protokoll for den 26. sesjon i Den blandete norsk–russiske fiskerikommisjon*, Oslo: Ministry of Fisheries, 1997, p. 2.
28. *Protokoll for den 28. sesjon i Den blandete norsk–russiske fiskerikommisjon*, Oslo: Ministry of Fisheries, 1999, Art. 5.1. The fact that Norway justified the high TAC by the alleged difficult economic condition of the north-west Russian population is not further discussed here. See Hønneland (2003, pp. 78–81) for an elaboration on the

argument that Norway's fishery authorities could avail themselves of a widespread Norwegian 'catastrophe discourse' pertaining to north-western Russia at the time to legitimize what some might characterize as poor fisheries management. Norwegian media painted a picture of an impoverished Russian region, whereas in fact it was one of the most affluent in the Federation. Furthermore, the high TAC would not benefit either the Russian state or the north-west Russian population in general, since nearly all cod was delivered abroad. In short, it served to make the north-west Russian *nouveaux riches* even richer.

29. *Nordlys*, 6 June 2001.
30. This breach with the harvest control rule met little criticism from the Norwegian public, quite unlike the situation a decade earlier when TACs had been set far above scientific recommendations (see Hønneland, 2003, pp. 54–7). The Norwegian fishery authorities' argument that the breach was justified by the exceptionally good condition of the cod stock seems to have been generally accepted. However, in the assessment of the Barents Sea fisheries management for a north-west Russian fishing company's application for Marine Stewardship Council (MSC) certification (in which the author of this book took part), non-compliance with the harvest control rule was criticized and led to a lower score on the indicator for harvest strategy: 'it cannot be verified whether the current harvest strategy, which has allowed departures from the plan, is precautionary. It is important that once a policy is agreed, it is adhered to as there are always pressures each year to depart from the agreement, which the rule itself should already be taking account of' (Southall et al., 2010, p. 88). Future compliance by the Joint Commission with the existing harvest control rule was identified as one of six conditions for the certification to be valid over time: 'The [Joint Commission] needs to apply the agreed rule and implement [it] over a number of years so that it can be evaluated in practice. If the rule is not meeting expectations, it can and should be revised and the new rule applied and <u>tested</u> in a consistent way. Arbitrary overriding of the agreed rule is not precautionary' (ibid., p. 64, underlining in original).
31. *ICES Advice 2010*, Book 3: *The Barents and the Norwegian Sea*, Copenhagen: International Council for the Exploration of the Sea, 2010, p. 6. Spawning stock biomass had been above the target reference point (460 000 tonnes) since 2002. Fishing mortality had been reduced from well above the limit reference point (0.74) around the turn of the millennium to well below the target reference point (0.40) since 2006.
32. The quota of 703 000 tonnes was in line with the advice from ICES; see *ICES Advice 2010*, Book 3: *The Barents and the Norwegian Sea*, Copenhagen: International Council for the Exploration of the Sea, 2010, p. 8. The original harvest control rule would have given a recommendation of 667 700 tonnes, up 10 per cent from 607 000 tonnes for 2010. If the quota could be changed by more than 10 per cent (a provision intended primarily to shield the fishing industry from drastic quota cuts), the recommendation would have been 896 000 tonnes. With the revised harvest control rule's provision that fishing mortality should not fall below 0.30, the recommendation – and the TAC – landed in between these two figures.
33. *Protokoll for den 29. sesjon i Den blandete norsk–russiske fiskerikommisjon*, Oslo: Ministry of Fisheries, 2000, Art. 13.2.1.
34. *Protokoll fra møte i Det permanente utvalg for forvaltnings- og kontrollspørsmål på fiskerisektoren i Henningsvær 16.–20. oktober 2000*, Bergen: Directorate of Fisheries, 2000, Art. 5.1.
35. *Protokoll for den 31. sesjon i Den blandete norsk–russiske fiskerikommisjon*, Oslo: Ministry of Fisheries, 2002, Art. 4.
36. Ibid.
37. *Protokoll for den 32. sesjon i Den blandete norsk–russiske fiskerikommisjon*, Oslo: Ministry of Fisheries, 2003, Art. 4.
38. *Protokoll for den 33. sesjon i Den blandete norsk–russiske fiskerikommisjon*, Oslo: Ministry of Fisheries, 2004, Art. 4.
39. Ibid.

40. Ibid., Art. 12.5.
41. *Protokoll for den 34. sesjon i Den blandete norsk–russiske fiskerikommisjon*, Oslo: Ministry of Fisheries, 2005, Art. 12.5.
42. *Protokoll for den 35. sesjon i Den blandete norsk–russiske fiskerikommisjon*, Oslo: Ministry of Fisheries, 2006, Art. 12.1.
43. Ibid.
44. *Protokoll for den 36. sesjon i Den blandete norsk–russiske fiskerikommisjon*, Oslo. Ministry of Fisheries, 2007, Art. 5.1.
45. *Protokoll for den 39. sesjon i Den blandete norsk–russiske fiskerikommisjon*, Oslo: Ministry of Fisheries, 2010, Art. 5.
46. The reports of the Norwegian Directorate of Fisheries about Russian catches in the Barents Sea from 2002 to 2008 are available on the Directorate's website www. fiskeridir.no.
47. *ICES Advice 2010*, Book 3: *The Barents and the Norwegian Sea*, Copenhagen: International Council for the Exploration of the Sea, 2010, p. 12.
48. This rough number figured in conversations with Russian fishery bureaucrats and in Norwegian media at the time. According to the protocol from the Joint Commission's session in 2006 – the only protocol during the 2000s where the *amounts* of overfishing are mentioned – the Russian side estimated the Russian overfishing in 2005 to have been 26 000 tonnes. Russia subsequently supplied ICES with estimates of Russian overfishing during the years 2002–07. According to these figures, overfishing ranged between 20 000 and 30 000 tonnes in the first three years of this period, peaking at 41 000 tonnes in 2005, before it was reduced to 28 000 tonnes in 2006 and 8757 tonnes in 2007. See *ICES Advice 2010*, Book 3: *The Barents and the Norwegian Sea*, Copenhagen: International Council for the Exploration of the Sea, 2010, p. 12. A further explication of these data is given in *ICES AFWG 2008*, Copenhagen: International Council for the Exploration of the Sea, 2008, p. 4. The main Russian argument to the effect that Norwegian estimates are too high seems to be the following: 'The Russian estimation takes into account that a considerable amount (57–58%) of resources fished in the Barents Sea (polar cod [a small cod fish related to, but separate from, North-East Arctic cod], Kamchatka crab) and Norwegian Sea (herring, blue whiting, mackerel, redfish) and, correspondingly, produce carried through [the Norwegian EEZ], are not cod and haddock [which they imply that the Norwegians assume].' The ICES Arctic Fisheries Working Group (AFWG) found no way to combine the two estimates and therefore undertook a double set of stock assessments and prognostic runs for cod and haddock. However, the Norwegian estimates were used as the foundation for quota advice.
49. *ICES Advice 2010*, Book 3: *The Barents and the Norwegian Sea*, Copenhagen: International Council for the Exploration of the Sea, 2010, p. 12. See also the reports of the Norwegian Directorate of Fisheries about Russian catches in the Barents Sea from 2002 to 2008 at www.fiskeridir.no.
50. See also *Statusrapport for 2008: Russisk uttak av nordøst arktisk torsk og hyse*, Bergen: Directorate of Fisheries, 2008.
51. In an interview with me in Bergen in May 2006, a leading figure in the Norwegian delegation to the Joint Commission said: 'the scientific cooperation has been the supporting pillar in the entire [bilateral management] cooperation [with the Russians]'.
52. The interviews with Norwegian scientists were carried out in Bergen during autumn 2007 and spring 2008 by my co-author Bente Aasjord.
53. The most infamous response to this policy was agronomist Trofim Lysenko's attack on (Western or 'unpatriotic') genetics. In line with the official view that a new breed of humans – the 'Soviet man' – could be created, he argued that acquired characteristics in a species could be inherited.
54. *On Necessity of Improvement of the Russian–Norwegian Management Strategy for Cod in the Fisheries in the Barents Sea*, Workshop for Discussion of the Joint Management of the Barents Sea Cod Stock, Nor-Fishing 2006, Moscow: VNIRO Publishing, 2006, p. 4 (emphasis added).

55. Interview in Moscow, December 2007.
56. Ibid.
57. I do not provide further technical details here, as this would have brought us well into the natural sciences, and is not necessary for my present argument. For such a presentation, see Aasjord and Hønneland (2008) and *ICES AFWC Report 2008*, Copenhagen: International Council for the Exploration of the Sea, 2008, Section 3.9.
58. PowerPoint presentation *North-East Arctic Cod Stock: State and Catch Forecast for 2008*, produced by B.N. Kotenev, D.A Vasilyev, O.A. Bulatov, V.M. Borisov, G.S. Moiseenko and E.N. Kuznetsova, presented to the author during interview in Moscow in December 2007.
59. See Aasjord and Hønneland (2008, pp. 300–301). The draft protocol is on file with my co-author Bente Aasjord.
60. *Fiskeribladet*, 23 May 2006.
61. *ICES Advice 2006*, Book 3: *The Barents and the Norwegian Seas*, Copenhagen: International Council for the Exploration of the Sea, 2006, p. 28.
62. Ibid.
63. Interview in Bergen, May 2006.
64. *Aftenposten*, 18 November 2002.
65. In an article about Russian implementation of international nature protection agreements, my co-author and I discussed the problem of getting Russian civil servants to attend international meetings: 'When it comes to international forums, such as conferences of the parties to a convention, Russian participation is low. Standard operational procedures limit official travel to a minimum. As one of our interviewees put it, "civil servants should not spend their time going on shopping trips abroad".' (Jørgensen and Hønneland, 2006, p. 15).
66. Interview in Kirkenes, June 2006.
67. Ibid.
68. Ibid.
69. In the early 1980s, Norway had tried to get the Soviet Union to agree on common minimum lengths for several fish stocks and mesh size in the Barents Sea, which implied that the Soviets would have to accept larger minimum length and mesh size than the requirements of their national legislation. See Chapter 3 for details.
70. Interview in Kirkenes, June 2006.
71. Ibid.
72. Interview in Bergen, May 2006.
73. The reason for this is probably that many more Norwegian than Russian vessels were involved in shrimp fishery.
74. He was referring to the 1999 quota agreement between Iceland, Norway and Russia, which put a stop to the 'Loophole disagreement' between Iceland on the one side and the two coastal states on the other (see Chapter 3).
75. *Murmanskiy vestnik*, 18 September 1999.
76. Interview in Kirkenes, June 2006.
77. This became a slogan in the Russian fishery press at the time. See, for instance, *Rybnaya stolitsa*, 15 November 1999: 'There was a meeting in the regional administration with participants from the fishing industry and scientists from PINRO where . . . the tactics and strategy of the Russian party [to the Joint Commission] were discussed. The principle that "ours" were to follow in the establishment of TACs for cod and haddock was adopted unanimously: not give in [to the Norwegians] on a single kilo.'
78. Admittedly, the solution chosen also solved some distribution problems at the national level in Norway.
79. He was also a member of 'the inner circle' of the Norwegian delegation to the Joint Commission.
80. Interview in Oslo, June 2006.
81. Direct translation of the informal Norwegian term for territory that is unfamiliar and far distant.

82. Preparatory meetings of one or two days between the delegation leaders and a small group of delegates from each side were introduced in the early 2000s. These meetings are normally organized one to two months before sessions in the Joint Commission, that is, in early autumn. In a way, the preparatory meetings replaced the summer sessions occasionally held during the 1990s, defined as an integral part of the session that 'commenced' the preceding autumn.

83. Interview in Bergen, May 2006.

84. There has traditionally been a high degree of stability in delegation leadership on both sides. The years around the turn of the millennium were an exception on the Russian side, with new delegation heads three years in a row. From 2001, stability returned with a Russian delegation leader serving for five years. Norway has had only three delegation leaders during the first thirty-five years of the Joint Commission: Gunnar Gundersen from 1976 to 1988, Gunnar Kjønnøy from 1989 to 1998 and Jørn Krog since 1999. The same pattern is visible in the Permanent Committee. Lisbeth Plassa was head of the Norwegian delegation from the Committee's establishment in 1993 to 2009, when Hanne Østgård took over. The Russians have changed their head of delegation far more frequently.

85. A top civil servant summarized the Norwegian policy in this respect: 'Norway is against bilateralization [of the scientific fisheries research in the Barents Sea]. ICES is the way, [the truth] and the life – international quality control and review'; interview in Bergen, May 2006. Russia is part of ICES, and official Russian policy is to remain so. However, actors in the Russian fishery complex routinely speak out against this multilateral research organization. For instance, at the celebration of the Joint Commission's thirtieth anniversary in 2006, a former Russian delegation leader gave a speech where he proposed that Norway and Russia should take care of fisheries research in the Barents Sea alone, without ICES interference. Russian criticism of ICES was particularly unconcealed during the difficult TAC negotiations around the turn of the millennium. From an article in the Norwegian fishery newspaper *Fiskeribladet*, 17 November 2000: 'Central participants in the Russian delegation [to the Joint Commission] show no mercy to the International Council for the Exploration of the Sea, ICES. While [the Joint Commission] will probably set a cod quota for next year that lies far above the scientific recommendations and not take into consideration the scientists' views on [sustainable] fishing patterns, the powerful shipowner and [State] duma representative Vladimir Gusenkov states that ICES is far from having the necessary credibility and objectivity. "ICES reminds me a bit of the scientific advice that we had during the Soviet era, where all recommendations were in accordance with the government's special interests. The difference is that ICES defends Norwegian interests in each and every way."' See also *Fiskeribladet*, 21 November 2000, where Gusenkov referred to ICES as 'an instrument in the hands of the Norwegian government'.

86. Interview in Oslo, June 2006.

87. For example, in a newspaper interview the leader of the enforcement section at the Norwegian Directorate of Fisheries, who has been a member of the Permanent Committee since its establishment, complained that the Russians were not willing to share satellite tracking data with the Norwegians, as they had promised in the Joint Commission that they would do: 'I have the impression that Russia doesn't prioritize this. I also think it's fair to say there's a lack of will on the Russian side. Instead of doing what we've agreed to do, there's unwillingness to implement measures' (*Nordlys*, 23 June 2006). A Norwegian public prosecutor left a meeting in the Joint Commission's newly established working group on overfishing and economic crime in protest: 'Walking out of the meeting was meant to mark my reaction to the fact that the Russian authorities and police had failed to meet as agreed, without offering any reasonable explanation. I saw no reason to be present if there would be no opportunity for a true exchange of useful information' (*Nordlys*, 7 June 2006).

88. This situation was not entirely new to the Norwegians. Also during 'the good years' of the 1990s, perceptions differed between the Norwegians and the Russians in the

Permanent Committee about what constituted sufficient control. The Russians were primarily concerned about 'paper control', that is, checking whether catch data submitted from vessels and shipowners were within allocated quotas. The Norwegians, on the other hand, constantly preached the need for *physical* control of the catch. At a Coast Guard seminar in the late 1990s, the Norwegians used an informal moment at a dinner party to give the Russians a set of scales as a present – in an attempt to get their message through in a more subtle and humorous way.

89.   Interview in Bergen, June 2011.
90.   Ibid.
91.   Ibid.
92.   Indications that a new wave of reform attempts was under way came when President Putin made his annual speech to the Federal Assembly in April 2007. For the first time, fisheries-related issues were given more than a passing mention in the President's address on the state of the nation. Putin called on the government to work out a system of measures to improve customs control and prevent overfishing; he argued for a halt to the practice of giving fishing quotas to foreign fishing companies, and spoke of the need to restore the country's shipbuilding industry (Jørgensen, 2009).
93.   Interview in Oslo, June 2006.
94.   Ibid.
95.   Obviously, it would not have set a good diplomatic tone for Norway to continue its unilateral estimates now that the joint method had been adopted.

5. Post-agreement bargaining at individual level

We now move away from the negotiation venues of the Joint Commission and its Permanent Committee to the fishing fields of the Barents Sea; from meetings between Norwegian and Russian scientists and civil servants to encounters between Norwegian Coast Guard inspectors and Russian fishers. These meetings take place on the open sea, when inspectors board the fishing vessels to check catches, fishing gear, holds and documentation, such as the catch log. All this takes several hours, so there is plenty of time to discuss matters related to the inspection as well as other things. There is also radio communication between Coast Guard vessels and the fishing fleet. Well into the 1990s, the Coast Guard had Russian interpreters on its staff, but in recent years the Russian captains' command of English has improved to such an extent that interpreters are considered unnecessary. The Coast Guard carries out fishery inspections on behalf of the Ministry of Fisheries, and inspection data are fed into the quota control performed by the Directorate of Fisheries. But the Coast Guard is more than a watch-dog: it is also responsible for search and rescue; it occasionally assists the fishing fleet in changing of crew, transport of material and ice-breaking; it can provide medical assistance and other services connected with the wellbeing of the seafaring community. In short, it is the state's representative in these vast areas[1] – it takes two days for a fishing vessel to get from Svalbard to the mainland.

This chapter provides accounts of the relationship between Norwegian Coast Guard inspectors and Russian fishers from two different sources: my own observations from the Coast Guard and interviews with Russian fishers, first in the late 1990s and then a second round of interviews ten years later (see the section 'Methodological considerations' in Chapter 1).[2] The main focus is on the dynamics between inspectors and fishers during inspection. What do the inspectors try to achieve, apart from the obvious checks of catch and documentation? How do they try to influence the fishers' behaviour? How do the fishers react? Do they adapt their behaviour? How is the work of the Norwegian Coast Guard generally perceived? I also bring in other topics from the interviews to place my main question in a context. For instance, when the fishers say something about their experience with the *Russian* enforcement authorities, this is probably their most important frame for comparison and should hence be treated as relevant for interpreting what they say about their perceptions of Norwegian inspectors as well. Without aiming to assess the general level of compliance in the Barents Sea fisheries, I occasionally refer to what my interviewees say about this – again, in order to place what they say about the Coast Guard in a context.

I use the generic term 'fisher' for any crew member of a fishing vessel. Obviously, inspectors from the Coast Guard are mostly in contact with the vessels' captains and other high-ranking members of the crew, such as those responsible for fishing operations, factory and holds. Most of my interviewees were captains or next in command, but some were rank members of the crew. The difficult question of who makes decisions on board a fishing vessel is largely left untouched here. It is widely assumed that the shipowners give general directions about fishing operations, but that the captains enjoy considerable room for manoeuvre. Captains, in turn, do not make their decisions in a vacuum. They might be influenced by interactions with the rest of the crew – and with outsiders such as Coast Guard inspectors.

Just as in Chapter 4, the present chapter is mainly an empirical account; we return to the theoretical discussion in Chapter 6.

AN OBSERVER'S ACCOUNT

The Norwegian Coast Guard patrols the areas of the Barents Sea where fishing activity is most intense. In winter and early spring, this is outside the coastline of North Norway, in summer and autumn mostly farther north, in the Svalbard Zone. Sometimes Coast Guard vessels shift between different fishing fields, inspecting a few vessels in one field before moving on to another one (and, as mentioned, they have other tasks than fishery inspections). Other times the enforcement vessels may lie in the same position for several days while the inspectors try to cover as many fishing vessels as possible. On occasion, the Coast Guard may use helicopters, so that inspectors can descend on fishing vessels for surprise inspections. Otherwise, inspectors approach the vessels via a small boat, and then climb up a ladder that has been lowered.

As mentioned in Chapter 1, I started my engagement in the Norwegian Coast Guard as interpreter and was also used as a witness at fishery inspections. Then I was trained as a fishery inspector and could conduct inspections on my own. During my five years in the Coast Guard, I was probably involved in around 150 inspections of Russian vessels, and a somewhat lower number of inspections of vessels from Norway and third countries. I would always be part of an inspection team, which normally consisted of two people (one inspector and one witness), sometimes more. Once on board the fishing vessel, we would usually be welcomed by the captain or another senior crew member and invited to the bridge or the captain's cabin to take a first look at the catch log and get some initial impressions of the vessel's recent fishing activities. We would normally board the vessels around half an hour before the trawl was supposed to be on deck; this allowed us to be present when the fish was unloaded from the net and spread on the deck. There we would make the necessary checks of the fish from the last haul and the fishing gear: measure the mesh size and the trawl's round straps, count the amount of by-catch (of other species than the target fish) and measure a sufficient amount of specimens of the target fish (sometimes the entire catch) to find out the proportion of the catch that was smaller than minimum allowable length. After catch and gear inspection, we would either continue our paperwork, take a look at the factory on board (where fish from the last haul would already be under processing) or go to the holds, where the processed fish was stored, normally in frozen form. Hold inspections were a real challenge: the inspection team had to calculate the exact amount of fish products on board. If the vessel had not started the trip recently, the hold would be filled up with boxes of finished fish products. We would make sample tests of the contents of some boxes: did the boxes marked with haddock really contain this species? Or was it the more valuable cod, concealed as haddock? Was it possible to assess the size of the fish (which depends on its degree of processing)? More importantly, the total number of boxes had to be counted – which could be really difficult if the hold was so full that there were boxes several metres below the surface, and if the hold had slanting walls, which was usual. Spot checks would also be made of the weight of the boxes: did they contain the indicated number of kilos? Paperwork would then continue on the bridge or, more often, in the captain's cabin. The inspection team would have to calculate the amount of fish products on board, multiply this by the relevant conversion factor (see Chapter 4) of the fish product in question, and see if the total amount corresponded to the amount of fish entered in the catch log (which is always done in round weight, i.e. the weight of the fish before it is processed). The inspection ended with the inspector and the captain signing the inspection form, which would conclude with either 'nothing to remark', an oral warning (for minor violations), a written warning (for more serious violations) or arrest (for very serious violations).[3]

It should follow from the above that an inspection was not a straight-forward affair. There was plenty of room for disagreement on how the revealed facts should be interpreted against the legal framework, and whether the figures arrived at were correct. The easiest part was document control. Russian catch logs were meticulously filled in; I never came across a Russian vessel with serious faults in the catch log.[4] My fellow inspectors – with more experience than myself from inspections of vessels from other countries – would often comment that nobody keeps as good logbooks as the Russians do. One recurrent theme of discussion, however, was the Norwegian requirement that data from each haul must be entered in the catch log immediately after the fish was moved from the deck to the factory, based on the captain's best judgement. It was then possible to make corrections after the fish had been processed. The Russians, for their part, would often wait until the fish had been processed, multiply the weight of fish products by the relevant conversion factor and then enter the necessary data in the catch log. The Russian fishers often argued that it was 'illogical' to make loose estimates when the exact amount could be supplied only a few hours later. The Norwegian inspectors could only explain that this was Norwegian law; the underlying principle was that actual catches should not be calculated 'backwards', from finished product to round weight. During inspection on deck, there would usually not be any argument as long as the inspectors found that the fish and the gear were in compliance with requirements. If the first check revealed smaller mesh size than allowed, the captain would ask the inspector to measure in another part of the trawl, which the inspector would then normally do. If the result was still negative, the captain might complain that the inspector had measured incorrectly, for instance by putting too little weight on the measuring device. Disagreement could then be sorted out by the inspector adding a piece of lead that gave the exact weight required by Norwegian law. If a sample count of by-catch revealed more than allowed, the inspector would count the entire catch – and similarly measure the entire catch if a spot check revealed too much undersized fish. On deck, it would nearly always be possible to 'reach agreement', though. If the entire catch had been counted and measured and found not in compliance with Norwegian by-catch and fish length regulations, there was little the captain could do. This did not result in a warning, however, as the caption could not be punished for having the 'wrong' kind of fish in just one haul. But it required him to change fishing position, which he would then normally do. Inspection of the hold involved more room for serious disagreement. If the inspection team calculated the actual catch on board as being significantly larger than what followed from the catch log (on which reports to the Norwegian Directorate of Fisheries were based), the captain would protest: 'this cannot be right'. Then the inspector would have to go through the whole operation once more, and often the captain would argue that it was impossible to count the exact number of boxes with processed fish in a more or less filled hold. It was necessary to rely on the information in the catch log. If the inspector maintained that there was a discrepancy between reported and actual catch, the result would be either a written warning or arrest. In the latter case, the fishing vessel would have to accompany the Coast Guard vessel to a Norwegian port, where the entire catch would be unloaded so that a more exact calculation could be performed. If the divergence was less severe, the inspection would end with a written warning to the captain. Then the captain would often complain that it is impossible to be 100 per cent correct and ask about the tolerance or flexibility allowed – which the inspector, for his part, was not allowed to reveal. Normally, he would reply that there was zero tolerance.[5]

So far, I have described the series of events that constituted an inspection, and have mentioned a few frequent topics of dispute between inspector and captain. However, inspections were more than that: they were also social events. Russian fishing vessels spent several months at sea without changing the crew, and I would assert that hospitality is a central aspect of Russian culture. When Norwegian inspection teams boarded Russian fishing vessels, then, the situation had a touch of 'a visit'. 'When you visited us last time . . .' the Russians would say, instead of 'during your last inspection'. These 'visits' were opportunities to bring out the best food (before my time, allegedly also the best drink), take a break from the ordinary routines, and exchange opinions, stories and souvenirs. We ate dinners and listened to the life stories of Russian captains, were shown pictures of their wives, children and home towns, and discussed all conceivable topics, from coffee prices to world politics and literary classics. The relationship between inspectors and captains was much more nuanced – and generally better – than one might expect between a watchdog and his objects of investigation. At the very least, this created an atmosphere in which I got the impression that the Russian fishers wanted to do their best to meet the requirements of the Norwegian Coast Guard – within the bounds set by Russian legislation. Yes, the Russians did refuse to sign the inspection forms in the Svalbard Zone (since Russia did not recognize this area as Norwegian; see Chapter 3), but that was often accompanied by an apologetic smile and a comment along the following lines: 'I would have liked to sign, but you know this is a case for our governments. You and me, we're both sailors and don't want conflict, but the politicians always find things to fight over.' And yes, the Russians did fish with nets of smaller mesh sizes than allowed by Norwegian rules in the Svalbard Zone (again: Russia did not recognize this area as Norwegian, so Russian fishers used the same trawl nets as in Russian waters, where smaller mesh size was allowed). Almost always, however, they changed to nets of Norwegian standard when requested to do so. (They would have such nets on board since they also fished in the Norwegian EEZ.) And apologies for not complying with Norwegian law in the first place (only with Russian law) were often accompanied by praise of the Coast Guard: one captain said he wished he could have signed, because without the Coast Guard there would have been no order in the Barents Sea.[6] The Coast Guard's distinct institutional identity among the Russian fishers was emphasized by the fact that they almost always referred to it by the Norwegian designation, *Kystvakt* (the –y pronounced as in the Cyrillic alphabet, as –u), instead of the Russian translation *Beregovaya okhrana. Kystvakt* was one thing, (their own) *Beregovaya okhrana* another.[7]

The efforts of the Coast Guard to avoid fishing in areas with much juvenile fish deserve special attention. In the Norwegian EEZ, the Directorate of Fisheries could close such areas, but in the Svalbard Zone – where the intermingling of fry was most prevalent – the Directorate could only *request* fishers to stay away (once again: owing to the jurisdictional disagreement about this ocean area). The Coast Guard would then board the fishing vessels in the area and present the captains with their calculations of the amount of small fish that was caught. The inspectors would try to convince the captains that continued fishing would do disproportionate harm to the stock, taking into account the limited value of small fish (although my impression was that the appeals were more aimed at the fishers' conscience and concern about the biological viability of the fish stocks than at their immediate economic prospects). Further, my impression was that compliance with such requests depended on how well founded the Coast Guard's argument was. Sometimes the evidence was overwhelming, and then compliance was quite immediate. Other times the Russians would ask for further test hauls or even present the Coast Guard with additional information from their own fishing activities, or from Russian research vessels in the area. For instance, they could agree that there was too much intermingling of undersized fish in an area, but only at specific fishing depths. In several instances, the Coast Guard modified its requests according to such information from the Russians. On rare occasions, the Russians would consistently disagree with the arguments of the Coast Guard. It was such an event that led to the first (albeit interrupted) arrest of a Russian vessel in the Svalbard Zone in 1998 (see Chapter 3).

The following extract from one of my own reports after a trip with a Coast Guard vessel illustrates a typical situation:

> Inspections in the Storfjord Trench concerned mainly by-catches of fry (cod, haddock, redfish). We undertook catch-control of 10 kg capelin on all vessels. The capelin was generally of good size, but on all vessels we found fry of the above-mentioned species.

The Russians repeatedly maintained that they had not had by-catches of fry previously; they also tended to 'confuse' cod with polar cod [a smaller fish related to, but distinct from, North-East Arctic cod]. On average, our inspections revealed between 10 and 15 fry.

On the morning of 7 November 1991 we were informed that the Directorate of Fisheries would request the Russians to stop capelin fishery in the area from N7600 to N7730, and from E1700 to E2200; further, that Sevryba [the former state union of fishery companies and organizations in north-western Russia] had already been advised of this by telex. Before receiving this signal [electronically transmitted message], we had inspected the MB-0129, *Polesye*, and had taken the opportunity to ask the captain how he thought the Russians would react to such a request. He replied that they would stop fishing in the area if they were so instructed from land, but that this could only take place after the weekend, since 7 and 8 November are national holidays in the Soviet Union (in connection with celebrations of the revolution, held for the last time this year). Until Monday the 11th, there would be no staff in the Sevryba offices who could take a stand on the telex from the Directorate of Fisheries.

Immediately before the capelin fleet's [radio] catch report at 1930B, I broadcast the request to the fleet. In return came several comments indicating that they could not understand that we had data that could justify closing off such a large area.

In the catch report right after this, the 'fleet chief' [the coordinator for the Russian vessels] said that he had contacted the authorities on land, and would inform the fleet as soon as he had received a response.

The 'fleet chief' was then aboard the MA-0060, *Kapitan Telov*; and at 2100B hours that same evening we went on board that vessel. At first he seemed rather terse, but gradually loosened up and made a highly sympathetic impression. When we had presented to him our calculations on how much fry was being taken by the entire fleet in the course of a day, he agreed that it would be unreasonable to continue fishing. He had sent a telex to Murmansk, informing them of the request from the Directorate of Fisheries, but explained to us that there would only be telegraph operators on duty, and that it would be difficult to get hold of higher-ranking staff because they were likely to be out, celebrating. However, he promised to phone the next morning, since there would be operative personnel in place then.

. . . I contacted the 'fleet chief' the next day. He informed me that he had been instructed to send a search vessel to the eastern part of the area in question, and let this vessel undertake a few hauls before the authorities would take a stand on the Norwegian request.

Later that same day we inspected one Russian and one Latvian vessel, and found a dramatic increase in the proportion of fry in the catches. (On the latter inspection we found 76 cod, 27 haddock, 115 redfish and 2 herrings in 10 kg of capelin.) The 'fleet chief' was immediately contacted and informed of the results of our latest inspections. We repeated our request that the fishery be halted at once. He promised to telephone the authorities on land once again. Shortly afterward, he called to tell us that all vessels would stop fishing at midnight.[8]

Social interactions could be important also in these situations. If the arguments of the Coast Guard were not immediately accepted by the fishers, inspectors might spend many hours on board fishing vessels explaining the rationale behind the request, calculating the amount of undersized fish being caught, arguing that it would be in the long-term interest of everybody to halt the fishery. The inspectors would dine with the Russians, relax over a cup of coffee, and sometimes even nap in between spells. And the Norwegian argument did not always prove to be the most well founded. Nor were the Norwegian authorities unwilling to take into account information from the Russian side. During one rather difficult inspection – the Russian captain disagreed sharply with the arguments of the Norwegian inspector, but, on the other hand, did not have to go far to get out of the closed area – it was decided, more or less in earnest, to resolve the disagreement through a game of chess. Arguments continued to cross the table throughout the game. The Russian won, and he stayed in the area. That same evening, the delimitation lines of the closed area were amended according to the argument of the Russian captain.

I argue that the Coast Guard is part of a seafaring community where actors have their clearly defined functional roles but to a large extent depend on each other and maintain social relations across functional lines. This sense of community is certainly not weakened by the fact that the interaction unfolds in the most spectacular Svalbard scenery and surrounded by a certain amount of polar romanticism. There is arguably a spontaneous spirit of community between people who earn their daily living in these isolated areas, under extreme climatic conditions. When inspector and fisher meet in the polar night and, over a cup of coffee, start talking about when the ice will come drifting in from the east, the situation is more reminiscent of a meeting between polar sea colleagues than of one between a watchdog and a potential criminal.

On a more critical note, the relationship between the Coast Guard and Russian fishers is not as close as that between the Coast Guard and Norwegian fishers. Language and cultural barriers are obviously present.[9] Also, my observations were done in the late 1980s and early 1990s, a period when relations with the Norwegian Coast Guard were possibly better than earlier, when the bilateral management regime was less mature, and Soviet fishers to a larger extent fished in other parts of the world, and better than they reportedly became from the late 1990s, when the Russian fishery authorities accused Norway of discriminating against Russian fishers in their enforcement activities. Finally, the reader might rightfully suspect me of remembering only the 'sunshine stories' from my time in the Barents Sea some twenty years later, or – even worse – of 'selecting in' what fits my theory and 'selecting out' what does not. To this, I can say that my observations preceded my preoccupation with compliance theory. From my very first moments in the Barents Sea, I was struck by the friendly atmosphere that characterized relations between fishers and inspectors, in particular the general willingness to listen respectfully to the other party's arguments. Towards the end of my time in the Coast Guard, though, I had reached master's level as a student of political science, and on the reading list was Jürgen Habermas's theory of communicative action, 'that type of speech in which participants thematize contested validity claims and attempt to vindicate or criticize them through argumentation' (Habermas, 1984, p. 18). Lifting my eyes from the books to the encounters between inspectors and fishers, I thought that this was communicative action – in action.[10]

RUSSIAN FISHERS' ACCOUNTS IN THE LATE 1990s

The most striking feature in my interview data from the late 1990s (see the section 'Methodological considerations' in Chapter 1) is the near-unanimous acclamation of the Norwegian Coast Guard as a desired and even necessary actor on the fishing grounds. Some interviewees reported disagreement with Coast Guard inspectors on specific issues. In particular, they expressed irritation at the fact that the inspectors did not accept the scales used on board Russian fishing vessels, but insisted on using their own (which sometimes gave different results). More generally, some said they viewed Norwegian enforcement efforts as a bit 'exaggerated' – or 'offensively high', as one captain expressed it. Another one argued that, compared to the Russian enforcement body, the Norwegian Coast Guard was excessively preoccupied with revealing violations and not sufficiently helpful in finding the fish:[11] 'Norway has decided to find all violators and punish them. It's not the right thing to do!'[12] Nevertheless, there was agreement within nearly the entire interview sample that the Coast Guard performed its duties to the general benefit of the Norwegian and Russian fishing communities. Most interviewees expressed the opinion that the Coast Guard performed a strict and just enforcement of regulations, and that its inspectors had an open and understanding attitude towards the fishers. When asked about their impression of Coast Guard inspectors, many interviewees started out with a general remark about the importance of personal qualities: 'Everything depends on human qualities; the first meeting reveals it all.' Then an overall positive impression of the Coast Guard inspectors was, in most cases, presented: 'I have never seen anything in the behaviour of a Norwegian inspector that contradicts common sense.' Predictability and integrity seemed to be valued. Typical words used to characterize inspectors were *kul'turnyy* ('cultured') and *gramotnyy* ('literate' or 'enlightened', in practice: well educated). The message seemed to be that the Norwegian inspectors were fair and incorruptible, as opposed to what Russian fishers were used to at home: 'They act "culturedly" [*kul'turno*], in a strict, but humane, fashion. They stick to the framework of the law.' Similarly, 'the inspectors don't go beyond their authority'. Further, the human quality of politeness was emphasized: 'They don't get coarse [*ne grubyat*].' Another fisher referred to the mutual respect of seamen: 'We have a good relationship. The inspectors are polite. We respect each other here at sea. We all have hard work.' One interviewee even placed his attitude to the Coast Guard in the context of a general respect for Norway as a northern nation:

> Norwegians are energetic, tough; they live under harsh conditions. To survive, you cannot tolerate the slightest deviation from the law. It's different in Africa, where bananas just fall down from the trees. That's why I respect Norwegian law. I understand that it's necessary [to comply with it]. I respect the people of Norway.

There seemed to be general agreement among the interviewees that the existing enforcement regime was sufficiently strict; surveillance was sufficiently extensive, inspection frequency sufficiently high and sanctions sufficiently severe. Some fishers even said they were uneasy about fishing in the Norwegian EEZ because of the 'strict regime' there. Often conditions in the Barents Sea were compared to the state of affairs in other ocean areas where Russian fishers were active: 'It's a very risky thing to cheat here in the north.' This fisher was comparing the north to the east – the Pacific Ocean or Russia's far eastern fishery basin. Some explicitly contrasted Norwegian order to Russian chaos: 'You have order; we don't.' Moreover, although a few, as indicated above, said the inspection frequency was too high, there seemed to be a distinct attitude among interviewees that the existing degree of enforcement was desirable: 'Control is necessary. There would have been more violations without control.' A few referred to morality as a possible explanation of compliant behaviour: 'The law is the law. It must be complied with.' Finally, most of my interviewees expressed respect for the role of marine science in the Barents Sea fisheries management:

> Of course, it's necessary to do research on the fish stocks. . . . The specialized research vessels do a very good job, but commercial vessels with researchers on board don't. Their main job is to fish. The specialized research vessels do what they're supposed to do. They're important, necessary . . .

Another fisher remarked:

> I have a positive attitude towards [marine scientists]. The Barents Sea is not inexhaustible. It's quite right that the quantity of fish is checked. The fish has to

be preserved for future generations. I'm not competent to assess whether their estimates are right; this is a specific science.

I have earlier subjected this interview material to interpretation using Rubin and Rubin's (1995) theories about 'cultural interviews' (Hønneland, 2000a, pp. 143–51). We recall from Chapter 1 that these authors single out various mediation forms by which information can be conveyed from interviewee to interviewer: A *narrative* gives facts about when, why and how a specific event took place, while a *story* additionally mediates some kind of moral or indication of the subject's worldview. *Myths* express important norms and values in the community to which the person belongs, while *accounts* are explanations that serve to justify a specific behaviour. A *front* represents the picture that subjects give of what they think is expected of them in the interview situation. Finally, *themes* are repeated descriptions of real or desired behaviour.

By and large, my Russian interviewees stressed *order* as the main characteristic of the Barents Sea fisheries. Some captains even said they feared fishing there, because of the strict enforcement regime in place. This was also reflected in the descriptions of Coast Guard inspectors as 'cultured' and 'competent'. The main *theme* of these interviews can be paraphrased as follows: 'We don't dare cheat. There is order in the Barents Sea fisheries. Norwegian inspectors are strict and incorruptible, but they are fair and behave in a cultured and enlightened fashion.' It might be appropriate to interpret the tendency to emphasize the effectiveness of the Norwegian Coast Guard inspectors as a *front*: the interviewees were praising the fellow countrymen (and former colleagues) of the interviewer. Or claims that inspection frequency was 'offensively high' can be treated as *accounts*, for instance as justifications that one must be allowed to cheat a little bit from time to time. There is, however, more to this. The notion that Russia lacks order (*poryadok*) and has to import it from the outside is a central myth in Russian history. It dates back to the legend of how Russia acquired its first emperor in the ninth century. According to mediaeval chronicles, the following message was sent to the Scandinavian Vikings: 'Our land is wide and powerful, and very fertile, but we have no order. Can you rule us?' This line of reasoning is recognizable in the nineteenth-century philosophical orientation of Westernizers and post-Cold War politicians alike. It is also reflected in everyday life in Russia. Most foreigners who have spent some time in the company of Russians will recognize similar statements: 'Oh, us Russians! We have all thinkable good qualities and the best of intentions, *but we have no order!*'[13] Hence, the interview statements can very well be interpreted as *stories*, representing central norms and values among the interviewees, or as widespread *myths* in the interviewees' culture. Myths have been repeated so often that they have been accepted as truth, and communicate a set of social expectations for behaviour. In this case, it is the expectation of Russian fishers that in Russian waters there is no order, while in the West there is.

In brief, there might be reason to take the Russian praise of the Norwegian Coast Guard with a pinch of salt. The interviewees might have been just polite or apologetic about their own shortcomings. Or they might have been expressing their own norms and values more than facts, or repeated widespread myths in Russian society. But – might it also be so that narrative dictates action? (See the observation on narrative theory in note 19 in Chapter 1.) May the expectation that, 'in the West, there is order' in fact influence the Russians' behaviour – law-abidingness in Norwegian waters, unrestrained cheating at home?

RUSSIAN FISHERS' ACCOUNTS TEN YEARS LATER

My more recent interview material is more diverse than my data from the 1990s, for several reasons. There are more interviews, and they were performed partly by me and my colleague in various Norwegian ports and partly by a Russian researcher in Murmansk (see Chapter 1 for details). However, the statements from the interviewees were rather similar, although there are also some tendencies pointing in different directions. Again, most fishers painted a picture of the Norwegian enforcement system as effective and Norwegian inspectors as strict, thorough and incorruptible. A few examples:

> The Norwegian system is rather effective; it works well. We're trying to get [our system] to where the Norwegians were more than twenty years ago.

> Their system, well, it's very effective. They have assembled all functions under the state, and everywhere you have to stand strictly to account. All catch logs and documents are stamped; you cannot make a replica. It's not like in Russia, where you can keep three books at the same time.

> The Norwegian control is effective because it's systematic and precise. They fight for their resources. Good job!

Some fishers painted a rather rosy picture of their relations with Norwegian inspectors. One said that the inspections take place 'in calm forms'. Another one:

> Top grade! They know their stuff and control without prejudice. They act with *Fingerspitzengefühl*. When the inspectors come on board, it's like they get a

hunch about how things are, right away. If they find anything [wrong], they behave in a 'stiff' and formal manner. But if everything is in order, they can talk with you, man to man, and drink tea with you.

More common in this round of interviews, however, were statements to the effect that Norwegian inspectors are too demanding, non-communicative and discriminatory. Such statements were also coupled to the image of Norwegian fisheries enforcement as 'effective', only now with the Russian fishers as 'victims' of an oppressive state policy. Nevertheless, some of these statements were also accompanied by conciliatory remarks, for example about the fact that strict systems produce good results, or that human qualities can qualify the strictness of the system. Here are a few examples:

> I don't know how the Norwegians monitor Norwegian fishers, but with us they've always been – how can I put it? – demanding. They usually act without wasting any words. And if they find anything, they won't listen to our explanations. They're only interested in their own work. For instance, they always examine just the catch log. What's in the vessel log doesn't interest them at all. I can't attempt to assess their work in a broader, global context, but they do work hard and that probably is of some value.

> The inspectors are very strict; you hardly dare crack a joke with them.

> They're strict in their control and are hard to please. Sometimes they do everything to catch an infringement – even the slightest minor point. Maybe their bosses have instructed them to do this, or maybe they're simply so thorough themselves – I don't know. But they carry out control strictly, and very often.

> They work well. The inspectors are always very exact. Sometime they can behave in an extremely formal way. For example, when they measure mesh size: if the mesh is even a tiny bit smaller than what's allowed, they immediately define it as a violation. That's probably the reason for the widespread view that it's hard to please the Norwegians. But it's individual. After all, the Norwegians are human beings just like us.

> Yes, the Norwegians work – but how? They pat their countrymen on the back, but try to keep us down.

Just as in my interviews in the 1990s, marine science was almost without exception referred to with respect. Typical labels were 'very good', 'very advanced', 'very necessary' and 'very serious stuff'. Several interviewees now – just as in the previous round of interviews – said they were not really competent to offer assessments: 'I'm not a scientist . . .', they would start. Russian inspectors, however, were presented in a bad light, and in very vivid language. One captain, asked about how effective he assessed the Russian enforcement system to be, exclaimed: 'It's not even worth asking such a question!' Some said they had never ever been inspected in the Russian EEZ, 'never even heard of anyone who has ever been'. Another noted:

> The [Russian] Border Guard spends little time at sea. You can check the weather by them: if they put out to sea, it means the weather is going to be nice; if a storm is coming in, they're the first to leave the place. For them, going out to sea is an event, something that could be written about in the newspapers and shown on TV.

Many complained about the lack of qualifications of Russian inspectors, ridiculing the fact that the Border Service is staffed by people with no fisheries competence but with experience from other branches of the FSB (of which the Border Service is part; see Chapter 3): 'Unfortunately, many inspectors cannot distinguish one species of fish from another, especially the military inspectors [from the Border Service]. . . . One of them is a colonel who used to be the conductor of a military orchestra!' Another fisher said about the inspectors: 'Fish they only know from restaurants!' Many referred to corruption among inspectors, and this was often woven into accounts of Russian chaos, 'Russian mentality' or 'Russian worldview'. As one fisher said: 'A Russian person does not like to comply with the law. Russians even write laws just to have something to violate' (laughs). Most respondents agreed that overfishing had taken place by Russian vessels, but claimed that it had now been reduced.[14] Most of them also maintained that overfishing was not a problem solely among Russian fishers. Asked whether overfishing in the Barents Sea was caused primarily by Russian fishers, one respondent exclaimed: 'What are you talking about?! The *norgs*[15] are also stealing! But they steal in the smart way, not like us: in the wild way.' The management collaboration between Norway and Russia in the Joint Commission was generally mentioned in positive terms. Several respondents spoke of Norway as a 'leader' in the collaboration with Russia. One of them said: 'I think Norway has worked very actively . . . been active in initiating [regulation and enforcement] arrangements. There hasn't been such an initiator in the [far] east. Japan hasn't been very active.'

As noted in the introduction to this section, there were many similarities between my interviews with Russian fishers in the late 1990s and ten years later. The big picture that was offered was one of Norwegian order and Russian chaos. As discussed in the previous section, we may rightfully wonder whether such statements are reflections of the interviewees' sincere experience of the outer world, or simply 'what people usually say' on these topics – reflecting *myths*, in Rubin and Rubin's (1995) terminology. I was a bit surprised at the force and vividness in the derogatory descriptions of Russian fishery inspections, coupled with the claim that Russian enforcement vessels were hardly ever put to sea. How could the fishers have such massive and apparently genuine experience of Russian fishery inspectors when they had barely met them, if at all? I raised this question with my Russian interviewer (see Chapter 1), who replied that she had been thinking along the same lines. She said something to the effect that 'this is how we Russians usually talk about our authorities, especially inspection bodies; we simply take for granted that they are incompetent and corrupt'.[16]

On this background, some caution should be exercised in treating my respondents' presentation of Norwegian enforcement as 'true' or genuine. Is it just a reflection of the myth that in the West there is order? If we, nevertheless, accept these statements as more or less sincere – if nothing else, they are the best indication we have about how Russian fishers experience Norwegian control in the Barents Sea – two main themes emerge. The first one goes approximately like this, similar to the main theme I identified in my interviews from the 1990s: 'Norwegian enforcement is effective; the inspectors are fair and incorrupt. They treat you with respect. If everything is in order, they even sit down to chat and drink tea with you, man to man.' Then there is a second theme, somewhat related, but with a different emphasis: 'Norwegian enforcement is effective, but it comes at a price. The inspectors are hard to please, and if they discover anything suspicious they don't want to listen to your arguments. You'd better stay on the right side of the law!' These two statements share the assessment of the Norwegian Coast Guard as effective. Implicitly, they also share the judgement that everything is fine as long as the inspectors do not find anything to put their finger on – but they are constantly searching to find exactly that. The second theme was far more prevalent in my second round of interviews than in the first one. This is not surprising, given the complaints presented by the Russian authorities and Russian media since the late 1990s that Norway is discriminating against Russian fishers in Barents Sea enforcement (see Chapter 4).

BARGAINING DYNAMICS

From my own observations in the Norwegian Coast Guard, I reported a good deal of communication – and some measure of bargaining – between inspectors and fishers in the Barents Sea. The prime example of effective bargaining, supported by extensive communication, was the Coast

Guard's endeavours to persuade Russian fishers to stop fishing in areas with too much intermingling of fry (primarily in the Svalbard Zone, an ocean area where coercive measures could not be employed). The main instrument was argumentation, at the scientific and practical levels. The inspectors appealed to the fishers' conscience, to perceptions of their own long-term interests, but most of all to their common sense. The fishers were urged to change fishing position not because continuing in the same place would have been a violation of the law, but simply because 'it doesn't make sense' to take up so much fry. And communication was two-way: the Coast Guard (with the Norwegian Directorate of Fisheries, which made the formal decisions on these issues) was ready to take the Russians' practical experience into account. If the Russian fishing or research vessels had more exhaustive data and more compelling arguments, for instance about intermingling of fry at different depths, the Norwegians would modify their initial request. The actual 'request areas' – the areas that fishers were requested to stay out of – thus emerged as the result of fact-based bargaining between the Norwegian enforcement body and the Russian fishing fleet.

There was not much room for actual bargaining during inspection, but there was plenty of communication, explanation and justification. The inspector was generally ready to explain the rationale behind a Norwegian rule and the procedures that he was following during inspection. If the first spot check of the mesh size indicated violation, he would be willing to measure again, if necessary adding a metal weight for control. If there proved to be too much undersized fish or by-catch of other species in a sample of the catch, he would measure or count the entire catch so that the exact percentage of intermingling could be identified. If discrepancies emerged between reported catch and the inspector's calculation of actual catch in the hold, he would go through his calculations once more, in the captain's presence. And the inspections were social events as well. As I have argued, communication was not limited to catch, documentation and fishing gear, but crossed over into the private sphere. More generally, there was a sense of community between those who had their occupation in these distant areas with tough climatic conditions. Further, the generally good atmosphere between the Norwegian enforcement body and the Russian fishing fleet made the fishers more inclined to follow requests from the inspectors, within the limits set by the Russian authorities.[17]

But there were, as indicated, limits to bargaining. As far as I observed, the inspectors would never deliberately ignore a violation, however minor it might be. All measurements of catch and gear were reported in the inspection form. So were smaller violations that were not directly connected to the fishing operations. For instance, if an incorrect flag was used, or there was a broken rung on the ladder that the inspectors used to board the fishing vessel, an oral warning would be given (and noted in the inspection form). The same goes for minor errors in filling out the logbook with no direct consequences for fisheries management. Nor would the inspectors give way in rather trifling requirements, such as the obligation to fill in last haul's catch in the catch log right away, based on the captain's best estimate, rather than wait until the catch had been processed to get a more exact weight. This was a source of much frustration among the Russian captains. There were limits not only to bargaining, but also to communication. The inspectors would not reveal the allowable tolerance between reported catch and actual catch on board, not even admit that there was such flexibility, much to the fishers' irritation.

Some of my interviews with Russian fishers confirmed my own observational data, but most of them painted a less positive picture of relations between Norwegian inspectors and the Russian fishing fleet. Most respondents characterized the inspectors as competent, fair and predictable, some of them even as reasonable, guided by common sense. My main impression, especially from the last round of interviews, was that the Norwegian inspectors were perceived as 'hard to please'. This seemed to involve a relentless search for violations ('they do everything to catch an infringement'), lack of flexibility ('if the mesh is even a tiny bit smaller than what's allowed, they immediately define it as a violation') and unwillingness to listen to the Russians' arguments ('they usually act without wasting any words – and if they find anything, they won't listen to our explanations'). Most annoying to the Russian fishers were obviously situations when the inspectors were not able – in the Russians' eyes – to give plausible explanations for the Norwegian rule (e.g. the rule that the weight of last haul's catch must be estimated right away, before it has been processed and a more exact weight can be established). The Norwegian enforcement system was generally characterized as effective, but, as many respondents claimed, 'that comes at a price'. The price here was obviously lack of flexibility. The situation was assessed differently by different individuals: some saw the stringent control as a problem ('It's not the right thing to do!'), others as a matter for emulation ('Good job!').[18]

So what is the situation on the Barents Sea fishing grounds actually like? Is there a brotherhood of polar seamen – or East–West antagonism and broken communication? There is no inconsistency here; I assume that both exist, or at least they have existed. There are good reasons to assume that my 'happy picture' of the Barents Sea fishing grounds in the late 1980s and early 1990s is less representative of the situation some twenty years on. Fisheries relations between Norway and Russia soured in the late 1990s, among other things after Norway for the first time attempted to arrest a Russian vessel in the Svalbard Zone in 1998, and for the first time carried out such an arrest three years later (see Chapter 3). According to the Russian media and Russian higher fishery officials (see Chapter 4), Russian fishers started to feel they were discriminated against by the Norwegian Coast Guard. This sentiment was confirmed in my interviews – it hardly featured in the first round of interviews (before the alleged discrimination started), but was quite prevalent in the second round. Furthermore, it is possible that the communication between the Norwegian Coast Guard and the Russian fishing fleet was more open and 'effective' when the Coast Guard had interpreters among its staff (also until the late 1990s). Russian captains are probably more proficient in English now than they were twenty years ago, but providing Coast Guard staff knowledgeable in Russian language and culture might well have been perceived by the Russian fishers as an extra service, one which both eased communication and made the fishers more accommodating to the Coast Guard's requests. As we have seen, lack of justification for specific rules was among the things that frustrated Russian fishers the most. At least, interpreters can facilitate communication in situations when such a need arises.

There is a great potential for fruitful communication in the perception among the fishers of a seafaring community, where people at sea respect each other across functional lines ('You and me, we're both sailors . . .'). This potential is utilized by the Norwegian Coast Guard in its efforts to make fishing vessels stay out of areas with too much intermingling of undersized fish. However, that seems to be where the bargaining ends. The Norwegian inspectors are trained to behave correctly and respectfully towards the fishing fleet, but they have no formal room for negotiation when it comes to violations that are uncovered. Nor did I ever see this taking place (surreptitiously or without legal foundation) in my own observations. I have argued elsewhere (Hønneland, 1993, pp. 93–8) that the Norwegian Coast Guard has a strong organizational culture that promotes meticulous inspection and reaction 'according to the book'. Letting one's personal judgement influence a decision is similarly an organizational taboo. Likewise, Norwegian fisheries legislation leaves little room for differing grades of punishment (see Chapter 2); even trifles are reported on the inspection form if the law so requires (although fines are increased for repeated violations). The unwillingness of inspectors to disclose the level of actual tolerance or leeway – the expressed denial that there even is such a thing – fills in the picture of an enforcement system with little flexibility, and minimal room for negotiation (at least during inspection).

BARGAINING RESULTS

Thus, bargaining results by the Norwegian Coast Guard on the Barents Sea fishing grounds are generally limited to reducing the fishing pressure on fry and other undersized fish, especially in the Svalbard Zone. But this achievement should not be underestimated. These northern areas of the Barents Sea contain large amounts of small fish, and the fishers would arguably have little incentive to stop fishing in such an area unless the intermingling of fry became so great that it totally overshadowed the 'usable' catch – which it hardly ever did. So if we do the same counterfactual move that we did in the last section of Chapter 4 and imagine the fishing grounds of the Barents Sea without the initiatives of the Norwegian Coast Guard to halt the fishing of undersized fish as early as possible, large amounts of fish would have been taken out of the sea long before they could be used for human consumption. Most likely, the small fish would simply have been thrown overboard, although this is illegal according to both Norwegian and Russian fisheries legislation (but this is a violation that it is possible for inspectors to discover only when they are actually on board a fishing vessel). I am not competent to evaluate the consequences in scientific terms, but this would obviously have contributed to unsustainable use of fish resources; it would also have interfered with the scientific models for stock assessment, since the discarded fish would not have been recorded in any statistics. Moreover, this part of the Coast Guard's enforcement activity is little known outside the Norwegian fisheries regulation system, since it seldom creates media-attractive episodes. In sum, it is a prime example of day-to-day bargaining between an enforcement body and fishing vessels.[19]

What of the room for bargaining during inspection? I have stated that there is no formal room for negotiation on the part of Norwegian inspectors, and concluded that also in practice this hardly ever takes place, supported as it is by an organizational culture that furthers discipline and 'doing things by the book'. I did report one example that might look like a bargaining situation, a rather informal one at that: the inspector who suggested resolving a dispute through a game of chess. It should be noted that this took place in the Svalbard Zone, where the Norwegian Coast Guard could not resort to coercive measures (or at least did not do so until the turn of the millennium). Moreover, the situation was rather atypical, and the inspector did not have much to lose (apart from his reputation as a 'serious' inspector), since arrest of the fishing vessel was not an option. But the episode gives us a glimpse of another counterfactual situation: what if the Coast Guard inspectors were given slightly freer rein? What if they had acted more as consultants than as policemen? (See Chapter 2.)

Might it contribute to greater compliance and more sustainable fishery if they could apply their own best judgement in each specific situation? After all, the inspectors, with their reputation as fair and competent, do enjoy considerable goodwill among the fishers. But would the bottom line be better? Or would too much be lost along the way?

NOTES

1. The Coast Guard was established together with the Norwegian EEZ in 1977, based on a previous, far less well-equipped, fisheries inspection service. The northern branch of the Coast Guard is headquartered at Sortland in Nordland County and has nine vessels at its disposal; the southern branch is in Bergen, with five vessels. Three new helicopter-carrying vessels (the *Nordkapp* class) were delivered in the early 1980s. They were the largest vessels of the Norwegian Navy and formed the core of the Coast Guard's northern branch until they were supplemented with two even more modern vessels after the turn of the millennium.

2. The section on observational data draws on Hønneland (1999b, 2000a), but gives a more extensive presentation than I have provided earlier.

3. As mentioned in the introduction to this chapter, it is not my aim here to assess the level of compliance in the Barents Sea fisheries, but it is generally high. My own investigations from the 1990s (Hønneland, 1998, 2000a) indicated that on average 12 per cent of inspections in the Norwegian EEZ and 15 per cent in the Svalbard Zone resulted in a reaction of some sort (oral or written warning or arrest; warnings for lacking reports in the Svalbard Zone not counted). Only 4 per cent of inspections in the Norwegian EEZ and 0.5 per cent in the Svalbard and the internal waters of Svalbard resulted in arrest. In 2010, 18 per cent of the Coast Guard's inspections led to reaction, 3 per cent to arrest; see *Aftenposten*, 27 December 2010.

4. However, as one of my interviewees noted (see below), the idea of keeping multiple logbooks is not alien to the Russians. The one presented to Norwegian inspectors would have to reflect actual catch on board, since the Coast Guard calculated the catch in the vessel's holds. In theory, they could use another catch log for presentation to the Russian authorities, who keep quota control of Russian vessels. Presumably this became more difficult when the Norwegian authorities started to forward data about landings from Russian vessels in Norway in 1993 (see Chapter 4), but only if the Russian authorities actually used the information from Norway in their quota control. According to my interviewees at the Norwegian Directorate of Fisheries (interview in Bergen, June 2011), there are indications that this was not the case.

5. If there was a tolerance, discrepancies within it were not documented, presumably so as not to reveal the level of latitude accepted. If, for instance, a (randomly selected) divergence of 3 per cent was allowed, revealed discrepancies of 1 or 2 per cent would not be documented in the inspection form, since the fishers would then understand that this was within the established level of tolerance. Instead, the inspectors would simply record that no discrepancy had been found.

6. I found this formulation in one of my own reports from the Coast Guard. This captain was probably not the only one to express such an opinion.

7. It was only after the Federal Border Service took over control in the Russian EEZ (see Chapter 3) that the term *Beregovaya okhrana* became used with some frequency.

8. *Tolkerapport (K/V Stålbas – 011191 – 141191)*, on file with the author.

9. In my interviews in the late 1990s, Norwegian fishers and fisher representatives reported similarly good impressions of the Coast Guard and its inspectors to those of my Russian interviewees (see next section). One fisher said: 'I have been in this profession

since the late 1960s, and it was us who wanted the Coast Guard; some things may be worthy of criticism, but we take it up with them and sort things out together.' And another one: '[The inspectors] are nice fellows; they sit and chat with us, they do.' To a larger extent than the Russians, the Norwegians would also express familiarity with the Coast Guard as an *organization*, its administration and leadership (Hønneland, 2000a, 2000b). One fisher representative said about the staff at the Coast Guard's northern headquarters at Sortland: 'I often ring them just to have a chat about conditions at sea.' And a fisher noted: 'We have a very good relationship with the base at Sortland. [The head of command] is a very fair chap.'

10. I went on to write my MA thesis about power and communication in the enforcement of the Barents Sea fisheries (Hønneland, 1993). At that time, I leaned more on 'big' Habermasian theory than on the emerging issue-specific literature on compliance in fisheries.

11. In Russia, both scientific and enforcement vessels have traditionally had a role in assisting the fishing fleet in locating promising concentrations of fish; see Hønneland (2004).

12. As mentioned in Chapter 1, I do not indicate where or when my interviews with Russian fishers took place, apart from the fact that it was in various Norwegian ports during 1997 and 1998. This is done to secure the anonymity of interviewees and those who helped me arrange the interviews.

13. I singled out 'order' as my 'key "key word"' in my own study of north-west Russian narrative and identity (Hønneland, 2010). It was the concept around which discussions about Russia and the West and about (Russian) north and (Russian) south evolved: north meant 'order', south 'disorder'; 'West' was synonymous with 'order', Russia with 'chaos'. Nancy Ries (1997) similarly claims that stories of calamities associated with living in Russia form a major speech genre in Russia. This is the story of Russia as a mythical land where everything is geared towards going wrong: a gigantic theme park of inconvenience, disintegration and chaos. 'You know what this country is, Nancy?', one of the interviewees asks her. 'This country is *Anti-Disneyland!*' (ibid., p. 42). It was the punchline of a conversation in which people traded examples of social chaos and absurdity in late-Soviet Russia. 'Our fairy-tale life' is another metaphor used by her respondents, referring to the monstrous political projects of the Soviet state (ibid., p. 43). Laments about Russian fairly-tale life typically ended with the following statement, 'Such a thing is only possible in one country – here, in Russia' (ibid., p. 49). However, Ries argues, 'Anti-Disneyland' also carries positive cultural value for Russians. Even when the 'Russia tales' had tragic elements, they were appreciated for their fascinating, amusing and astonishing epic. They made people feel personally part of the intense Russian drama (pp. 49–50). Russians may feel ashamed of their country when they are abroad, but, as one of her respondents puts it, 'within themselves they are all very proud that they are Russians, that they come from such a country, which has such a strange history' (ibid., p. 50).

14. While we cannot exclude the possibility that my interviewees were in fact well informed about the ups and downs of Russian overfishing at the aggregate level (which I actually doubt they were, given the limited attention to the issue in the Russian public), the similarity between their stories might in fact reflect narrative convention, rather than reality. Gergen (2001, pp. 254–5) claims that, in order to maintain intelligibility in the culture, the story one tells must employ the commonly accepted rules of narrative construction. As an empirical example of how life stories are constructed, he refers to how American adolescents characterize their life stories according to narrative conventions – happy at an early stage, difficult during the adolescent years, but now on an upward swing – conventions that do not necessarily reflect actual events in their lives, or their perception of them: 'In these accounts there is a sense in which narrative form largely dictates memory. Life events don't seem to influence the selection of the story form; to a large degree it is the narrative form that sets the grounds for which events count as important' (ibid., p. 255).

15. The usual Russian word for Norwegian is *norvezhets*. *Norg* is slang for Norwegian, slightly derogatory but very common in north-western Russia in recent years.

16. Again, this can be a case of narrative convention, rather than personal experience, dictating how an issue is presented (see above).

17. This implied that they could change nets to satisfy Norwegian mesh-size requirements even in the Svalbard Zone (which was, strictly speaking, not against Russian law), but they would not sign the inspections forms there, since that would not have been acceptable to the Russian authorities.

18. Those who saw it as a case for emulation could, for instance, see the Norwegian enforcement system as contributing to the long-term interest of the Norwegian and Russian fishing communities, which was a good thing even though – as an unavoidable side-effect of an effective enforcement system – some fishers had their rights slightly infringed here and now.

19. In the previous section, I raised the question of whether my observational data paint an overly rosy picture of the situation on the Barents Sea fishing grounds. By this, I did not imply that this type of bargaining between the Norwegian Coast Guard and the Russian fishing fleet ended in the 1990s – only that the conditions for successful bargaining became more difficult.

6.   Conclusions

In Chapter 1, I posed two theoretical questions: why do people obey the law, and why do states abide by their international commitments? We have seen that 'formal' models of compliance to a large extent presuppose unitary, rationally calculating actors driven by self-interest, and a social logic that will necessarily follow: a crime being committed, a common-pool resource destroyed, an international treaty concluded and subsequently complied with. Empirically, these models are used to study how self-interest, deterrence and power play out in real-world situations. 'Enriched' models of compliance assume that actor motivations are more mixed and social dynamics less stylized and predictable; and research efforts have focused on how norms, legitimacy and institutional organization affect compliance. The theory of post-agreement bargaining narrows in on how states promote the compliance of other states through interstate communication after a treaty has been concluded – unlike traditional studies of implementation and compliance, which have been more preoccupied with activities at the national level.

In this concluding chapter, I start with a brief summary of Norwegian bargaining experiences with Russia in the Barents Sea fisheries, at the state level and the individual level. The exact levels of compliance have not been in focus in this book, but my empirical accounts allow for rather robust conclusions on this issue as well, at least at the state level.[1] In the following two sections, I ask why Russia complies with its international commitments in the Barents Sea (when it does so) and why Russian fishers comply with Norwegian law (to the extent that they do). Finally, I revisit the concept of post-agreement bargaining.[2]

BARGAINING EXPERIENCES

A good amount of post-agreement bargaining has taken place between Norway and Russia concerning the management of the Barents Sea fisheries, mostly on the Norwegian initiative. Norway has tried to get Russia to take overfishing seriously on two occasions, first in the early 1990s and then in the mid-2000s. On the first occasion, the Russian side was quickly convinced that overfishing did in fact constitute a problem, and entered into new collaborative arrangements with Norway in the enforcement of the Barents Sea fisheries. This mainly implied the exchange of catch and landing data. In the 2000s, however, the Russian response was lukewarm. Initially, the Russians were interested in investigating possible overfishing, but this interest vanished when it became clear that Norway defined this as more or less exclusively a Russian problem. Nevertheless, agreement was reached in 2009 on a joint method for assessing the total catch from the Barents Sea. In the meantime, the 2007 NEAFC port state control regime had largely solved the problem of overfishing in the Barents Sea, although Russia has remained unwilling to present the basic documentation about Russian trans-shipments to Norway, and there is uncertainty about Russian readiness to prosecute violators. Norway had more success in getting the Russians on board when it came to the coordination of technical regulation measures, the joint introduction of new regulations and the 'automatization' of the setting of the TAC. As I concluded in Chapter 4, we cannot be absolutely certain that these decisions were 'right', but at least they brought the management of the Barents Sea fish resources closer to the standards around which international science and politics converge. This was possible because Norway did not simply leave implementation of Russia's international commitments to Russia itself, but engaged actively in post-agreement bargaining.

As a point of departure, one might expect such bargaining to take place between the parties 'over the table' – in this case, at plenary sessions of the Joint Commission. In practice, I have identified two other main tracks of Norwegian negotiation efforts: from bargaining at lower levels to approval by the Commission; and bargaining by the two heads of delegation, with decisions subsequently anchored in the respective delegations. Many issues have been negotiated and agreed upon in the Permanent Committee and its sub-groups before being presented to the Commission for final approval. This was the case with the establishment of enforcement collaboration in 1993, the harmonization of technical regulations (such as conversion factors and procedures for closing and opening fishing grounds) and joint introduction of new regulatory measures (like selection grids and satellite tracking) throughout the 1990s, as well as with more recent initiatives like the joint method for estimating total outtake of fish from 2009. In these cases, the challenge of reaching agreement between the two states was in practice handed over to technical experts (civil servants at lower levels or scientists). If the role of the Commission was not formally reduced to rubberstamping, in practice at least the agreements reached at lower levels were routinely accepted by the Commission. In a somewhat related manner, the established scientific collaboration between

PINRO and the Norwegian Institute of Marine Research functioned as a buffer against the introduction of the new Russian methods for estimation of fish stocks that were advocated by the federal research institute VNIRO but did not meet ICES standards for precautionary fisheries management. Here fundamental agreement on scientific principles had evolved over many years between Norwegian and Russian scientists under the auspices of ICES. Norway had intensified its support to PINRO, including financially, after the break-up of the Soviet Union. Whereas the Norwegian intentions were more altruistic – including Russia in the international scientific community – this investment could be 'cashed in' by leading Russian scientists showing support in the Joint Commission for Norway's position on new methods of stock assessment. But there were also risks associated with efforts to influence the lower levels of the Russian bureaucracy. The ease with which the Permanent Committee reached agreement allegedly led to suspicion in Moscow: were these scientists and civil servants defending Russian interests, or were they becoming too friendly with the Norwegians? Similarly, PINRO was, at least indirectly, suspected of running the errand of Western interests, and found itself squeezed financially and challenged scientifically by VNIRO.

The other main track of argumentation that I identified was direct communication between the two heads of delegation – mostly with their respective interpreters, or sometimes just the two of them, and on occasion in the Commission's 'inner circle'. The TAC has always been handled at this level, and not in plenary sessions. The same goes for many other important decisions, such as the introduction of new procedures – although, as we have seen, some new procedures were introduced through agreement at lower levels and then approved by the Commission.[3] We saw in Chapter 4 how the Norwegians around the turn of the millennium worked consistently to prepare the ground for the 2002 harvest control rule. First, they yielded rather a lot in the difficult negotiations in 1999 in order to 'keep the Russians happy'.[4] Next, they got the Russians on board with a three-year quota in 2000, an arrangement that included elements of the ensuing harvest control rule. And we saw how, in the final stages before the harvest control rule was adopted, the Norwegian delegation leader and the Norwegian Director of Fisheries – a member of the 'inner circle' and most likely the true father of the harvest control rule – 'worked on' the Russian delegation leader, first at a preparatory meeting and then in the Commission itself, to get him to accept the new rule. Once the rule was adopted, the head of the Russian delegation credited it to his own scientists, presumably to reduce any impression of the harvest control rule as being a Norwegian invention.

My Norwegian interviewees, who were high-ranking members of

the Norwegian delegation to the Joint Commission and the Permanent Committee, agreed that Norway had been the leading force in the collaboration, at least after the break-up of the Soviet Union. ('We have always been the proactive ones. The initiatives have always come from the Norwegian side. We've taken up the things we felt were wrong.') As a result, they saw the need to create ownership of the proposed measures on the Russian side. This was done by meticulous and persistent arguments (no short cuts), and by taking things 'in several rounds', from lower levels to the Commission itself. The introduction of selection grids was an example of a step-by-step process that gradually bound the Russians, if not formally, then in practice. First, selection grids were introduced in the shrimp fishery, which was mainly a nuisance to the Norwegian fishers, since they were more involved in that fishery than the Russians were. This, however, sparked the interest of Russian scientists and technical experts in the grid technology, and talks ensued about the possible use of grids also in the cod fishery. Practical exploration of the technology followed. By the time the technical experts had agreed first in a sub-group to the Permanent Committee and then in the Committee itself, the Russians had allegedly 'come too far' to pull out. This turn of events might have been unintended on the Norwegian side, but it serves to fill in the picture of negotiation dynamics in the Norwegian–Russian fisheries relations.

My interviewees described the negotiation atmosphere as fundamentally good – 'open and relaxed', as expressed by the current head of the Norwegian delegation to the Joint Commission. A prominent member of the Norwegian delegation to the Permanent Committee characterized working relations in the Committee as generally amicable: 'Even though we've hit them on a sore point, I have never heard one single unpleasant word.' But relations at this level deteriorated after the turn of the millennium, apace with the Norwegian documentation of Russian overfishing. As seen from the Norwegian side, the obstacle was the Russian civilian enforcement body, subordinate to the Federal Fisheries Agency:[5] 'it's like sitting in a rowing boat: as long as we're together, we row in the same direction; but when our ways part, they row like mad in the opposite direction'. This agency even withheld from their own colleagues in the Russian Federal Border Service data about landings from Russian vessels in Norway, received from the Norwegian Directorate of Fisheries according to the data exchange scheme established in 1993. The Joint Commission in 2009 agreed on common Norwegian–Russian procedures for estimating total catches from the Barents Sea, and there is no longer any documented overfishing there, but working relations have remained difficult in the sub-groups of the Joint Commission and the Permanent Committee on enforcement and economic crime: 'There's been a bit of a tug-of-war and bad atmosphere', as one interviewee put it.

At the level of individual bargaining between Norwegian inspectors and Russian fishers on the Barents Sea fishing grounds, I concluded in Chapter 5 that the Norwegian Coast Guard's efforts to get Russian fishers to desist from fishing in areas with too much intermingling of fry contributed to the sustainability of these fisheries. Especially in the Svalbard Zone – where deterrence could not be effective owing to the lack of sanctions in the event of revealed violation – this achievement could be attributed to communication, or bargaining. I showed in Chapter 5 how the Coast Guard spent considerable time trying to convince the Russian fishers, presenting data about the harm that would be done to the fish stocks if fishing operations were not discontinued. I also showed how the Coast Guard was ready to listen to arguments from the Russian side, and adjust its requests according to data provided by the Russians. Hence, the actual 'request areas' emerged as the result of bargaining between the Norwegian enforcement body and the Russian fishing fleet. During actual inspection, however, there was less room for negotiation. The Norwegian inspectors were not allowed to depart from the detailed requirements set by Norwegian law, and in practice they seldom (if ever) did so. The Russian fishers I interviewed generally presented the Norwegian inspectors as 'hard to please', complaining about lack of flexibility and unwillingness to listen to the Russians' arguments. On the other hand, the inspectors were characterized as fair and competent, sometimes even ready to talk with you 'man to man'.

WHY DOES RUSSIA COMPLY?

I concluded above that Russian compliance with a number of the country's international commitments in the Barents Sea fisheries was the result of Norwegian post-agreement bargaining.[6] This claim does not, however, fully explain *why* Russia complied. Did Norwegian bargaining efforts help Russia to conclude that adherence to the relevant international obligations was indeed in Russia's own interest? Were there some common underlying norms invoked in the bargaining process? Or were there institutional features in the bilateral management regime that favoured compliance?

We saw in Chapter 2 that traditional realists (classical and offensive realists) hold that states comply with their international commitments since these commitments reflect their interest; the underlying assumption is that states enter into only those agreements that can help to further their (military or economic) interest. Can this point of departure help us understand

Russian compliance with international fisheries agreements in the Barents Sea? Let us take the UN Fish Stocks Agreement as an example. A central feature of this agreement is arguably that it introduced the precautionary approach in international fisheries law. Russia is party to the Fish Stocks Agreement, but the precautionary approach is systematically shunned in Russian fisheries law, and seems to enjoy no legitimacy in the Russian fisheries bureaucracy. As we saw in Chapter 4, the director of the Russian federal fisheries research institute VNIRO claimed in an interview that the Fish Stocks Agreement was 'written by Greenpeace with money from the CIA', designed to harm Russian interests. I will not go into a discussion of why Russia signed and ratified the Fish Stocks Agreement in the first place, but it is problematic to argue that the Joint Commission's gradual adaptation to the precautionary approach as such, and to ICES's operationalization of the principle through the precautionary reference points, can be explained by Russia adapting its behaviour to its declared best interest (other than the possible interest in appearing cooperative towards other states). Nor is it plausible to argue that Norway has exerted military or economic power over Russia to make the latter comply. Norway may in some respects be categorized as an economic great power (and in fact did support the Russian fisheries bureaucracy financially during the 1990s), but Russia has considerable economic clout and is still a military great power. If military or economic capabilities were decisive, one would expect Russia to press through its primary preferences, and not seek compromise or even accept solutions explicitly against its will.[7]

Defensive realists and neoclassical realists are less categorical in their assumptions about the relationship between interest and compliance, and prone to view compliance as either more 'accidental' or the result of more diverse factors. As we saw in Chapter 2, defensive realists (Walt, 1998) claim that security is more abundant than assumed by offensive realists (Mearsheimer, 2001), so that states can be more relaxed in their pursuit of relative advantages. As a consequence, compliance can follow also when this is not in the immediate interest of the state in question. Neoclassical realists (Wendt, 1992) emphasize that a state's environment is murky and difficult to read, and includes both international and domestic constraints. The environment must be interpreted as states 'move along', so decisions about compliance may be rather 'accidental' or not always 'right' (in the sense of corresponding to the state's interests). In our case, Russian compliance with international fisheries commitments can – as I have already argued – hardly be explained by the immediate concurrence between one overarching material Russian interest and a specific compliant behaviour. For one thing, 'the Russian interest' is no unequivocal entity. Various different interest groups have been involved in fisheries management

cooperation with Norway: scientists, technical experts, the fishing industry and civil servants at many levels. If states are less coordinated in their
foreign policy than is often assumed (Allison and Zelikow, 1999), this
has certainly been the case with post-Soviet Russia. Institutional conflict
has characterized Russian fisheries management since the early 1990s
(Hønneland, 2004), and the Russian bureaucracy (like its Soviet predecessor, for that matter) has traditionally been strictly compartmentalized. As
a glaring example, there was no contact between the two Russian fisheries
enforcement bodies until they were 'forced together' in the Permanent
Committee, together with the corresponding bodies of governance from
the Norwegian side.[8] We have also seen that there were strong disagreements between the north-western fisheries research institute PINRO and
the federal fisheries research institute VNIRO. Thus, it is hardly surprising that the aggregate 'Russian interest' was difficult to identify for
the decision-making authority in our case, for instance the head of the
Russian delegation to the Joint Commission. It might have been difficult
to interpret the external environment, and different domestic interest
groups had their respective agendas. It might have been rather accidental
who got his way – or at least more the result of a domestic power struggle,
or corruption,[9] than of Russia as a state assessing its best interest. Another
example of a decision with which Russian dissatisfaction was widespread
in hindsight (cf. the discussion about the precautionary approach above)
concerned the introduction of mandatory selection grids in the Barents
Sea fisheries. We have noted various factors that can explain why Russia
accepted this move in the Joint Commission in the late 1990s, despite
the general (later) verdict by Russian fisheries actors that this was not in
Russia's interest. Russian technical experts had become excited about the
grid technology, and when the issue reached the Joint Commission for
final approval the Russians had by then 'come too far' to pull out. This
approach opens the way for the possibility that a state may enter into an
agreement (and later comply with the agreement) that is not in its strictly
defined interest. However, it does not say anything about the mechanisms
through which this takes place.

How about normative theory? We saw in Chapter 2 that Franck (1990)
claims that state compliance with international law generally follows from
the legitimacy of the specific rule. Have the Russians adapted their position
and politics in the Joint Commission mainly because they have deemed the
Norwegian initiatives legitimate? There are many indications that this is
not so. We have already seen that both the precautionary approach (which
stands out as the 'heading' of all Norwegian initiatives in the bilateral
fisheries management regime since the mid-1990s) and the introduction
of selection grids (which is an example of a specific management measure)

were widely rejected in Russian fisheries circles after they were adopted and implemented. Moreover, it is generally argued in Russia that ICES stock estimates are deflated and that the TACs in the Barents Sea could have been higher ('precaution has gone too far'), and there is widespread displeasure with the restraints imposed by the harvest control rule. We have also seen that the Russian side rejected the Norwegian initiative to document total catches (to reveal overfishing) after the turn of the century, and Norwegian assessments of Russian overfishing were routinely dismissed. Admittedly, some of the derogatory statements about the Norwegian initiatives might have aimed at pleasing specific interest groups in Russia (or reinforcing Russian identity), but more positive judgements have rarely been voiced, even in direct communication with the Norwegians.[10] So, legitimate solutions? Hardly. The Russians have continued to accept the Norwegian initiatives (although more falteringly since the turn of the millennium than in the 1990s), but then they routinely dismiss them after the fact. Why?

If the Russians do not view the Norwegian proposals as particularly legitimate, there is one other possible normative explanation: there exists a normative community between these two coastal states in the Barents Sea, and this furthers the development of agreement between them on specific issues. Geographical proximity is one thing; shared functional dependence – and tradition – is something else. We recall how the current head of the Norwegian delegation to the Joint Commission reflected: 'seamen are used to helping each other when the need arises, and we see this in the Commission as well. And our neighbourly relations are important . . . . There is an underlying recognition of the closeness of reality.' What he was hinting at was – in addition to the neighbourly relations – the norms of the *seafaring community*. One central seafaring norm, he said, was that people at sea help each other when this is necessary – between the lines: even if this is not in one's own interest, and without regard to nationality. He said outright that this spilled over from the fishing grounds to the negotiation venues of the Joint Commission. Most Commission members, both Norwegian and Russian (but especially the Russians),[11] have personal experience from the sea, where they can be expected to have adopted traditional seafaring norms, such as that of solidarity among seafarers of all nationalities. So when the Joint Commission convenes, *the meeting is not just between representatives of two different states, but also one between 'seafarers' from different nations*.[12] This arguably makes the parties to the Commission more disposed towards compromise, national interest and legitimacy aside.

This brings us one step closer to a possible understanding of the dynamics that have led Russia to accept several Norwegian initiatives that it at the same time openly dismisses. But we still have one major perspective on state compliance with international law: the institutional perspective. As we saw in Chapter 2, institutionalists maintain that the organizational form of an international arrangement, or regime, has a separate influence on state compliance, often formulated as 'Institutions matter'. Non-compliance, it is claimed, may be the result of institutional imperfection, in the international regime itself or its member states. For instance, states may have limited capacity to carry out their international obligations, in which case technical or financial assistance can be provided by other states, or an international organization, in order to enhance compliance. In the Barents Sea fisheries management, this was what Norway did in the 1990s (and to some extent also later). Russia was perceived to be lying there 'with a broken back', and Norway came to the rescue. This included the provision of landings data from Russian vessels in Norway, technical and financial aid to PINRO and Murmanrybvod, and reimbursement of all expenses in connection with meetings in the Permanent Committee, its sub-groups and the joint seminars for fishery inspectors. Did this improve compliance? Well, the landing data were allegedly never used, and Murmanrybvod and its institutional successors developed into the most fervent opponents of Norwegian precautionary initiatives ('row[ing] like mad in the opposite direction'). The joint inspector seminars may have had some long-term effect; for instance, young inspectors who attended these seminars in the 1990s later climbed to higher positions in their respective fisheries hierarchy, which laid the foundations for potentially fruitful collaboration at these levels. For example, one inspector later became head of the north-west Russian fishery inspection branch of the Federal Border Service and, according to my Norwegian interviewees, contributed significantly to the productive collaboration between this organization and the Norwegian enforcement bodies in the mid-2000s – until he was ousted (possibly for being perceived as overly collaborative with the Norwegians). By and large, however, the technical and financial assistance to Russia in the 1990s ricocheted on Norway, with Russian charges that Norway was trying to 'invade' the Russian fishery bureaucracy with dubious motives. The most successful assistance initiative was arguably that which was provided to PINRO, but as we have seen this hit back on PINRO itself, with financial constraints from Moscow and allegations of unpatriotic work. In sum, the assistance might have had some effect, but that came at a price. It can hardly be considered as a significant contribution to explaining Russian compliance.

Further, institutionalists argue that *bureaucratic processes* often favour compliance over non-compliance. Bureaucratic capacity does not normally allow for weighing the costs and benefits of compliance versus non-compliance in every specific case, so some kind of *standard operating procedure* is usually established, deliberately or 'by accident'. We recall that Chayes and Chayes (1995) accentuated *iteration* in itself: *compliance is sometimes the result of inter-state bureaucratic procedures established through iteration over time.* If compliance in itself has not become standard operating procedure, there has been a *drive towards agreement,* or *towards compromise,* in the Joint Commission – visible since its establishment in the mid-1970s, but accelerating with time. The foundation for the bilateral management regime was the 1975 agreement that split cod and haddock quotas 50–50 between the two parties. Then followed the sharing of the capelin quota (60–40 in Norway's favour) a few years later, and the mutually beneficial quota exchange arrangements between the parties (granting extra cod and haddock to Norway in return for some exclusively Norwegian species to the Soviet Union) from the late 1970s to the early 1990s. With the 1990s came a rush of new agreements on enforcement and technical regulation, negotiated in the Permanent Committee and its sub-groups and subsequently adopted by the Joint Commission. The pace slackened a bit after the turn of the millennium – but agreement was now reached on far more important issues, notably the 'automatization' of the TAC setting, in accordance with ICES precautionary reference points. After some difficult years with Norwegian allegations and Russian denials (in part, at least) of Russian overfishing, compromise was reached on a joint method for estimating overfishing in 2009. The same year the parties agreed to treat Greenland halibut as the fourth joint stock in the Barents Sea, after several years of scientific exploration and political bargaining. Furthermore, in that year the Commission also resolved one of its longest outstanding issues: joint minimum allowable mesh size and fish length for the entire Barents Sea – again through compromise. On top of this – and with far wider implications – came the agreement between Norway and Russia on the delimitation of the Barents Sea in 2010 (see Chapter 3). Among the Norwegian public, the agreement was generally hailed as an expression of the ability of the two neighbouring states to reach compromise.[13] I venture the assertion that reaching agreement has become a separate goal in the fisheries relations between Norway and Russia, something similar to an *institutional guideline* for the Joint Commission and its sub-groups. At the various levels of the collaboration, there is awareness that the higher levels expect agreement to emerge. Therefore the focus is on the room for compromise. In the Joint Commission itself, the parties take pride in their ability to reach agreement, expressed as 'no task is too big for us'. In a way, *compromise has become the modus operandi of the Joint Commission.*

Another related but somewhat different example of how institutional factors have affected agreement (and indirectly Russian compliance with the precautionary approach) was the gradual introduction of selection grids. In this case, agreement between Norway and Russia was reached not only because of the general 'rush towards compromise', but more specifically because the established procedure gradually came to bind the parties. It had become customary since 1993 to let the Permanent Committee, and soon also sub-groups appointed by the Committee, explore possible new matters for collaboration between the parties. As we saw in Chapter 4, the coordination of established regulatory measures and the introduction of new ones often took several years, from the first initiative until the measure was formally adopted by the Joint Commission. During this time, the technical experts in the sub-groups routinely reported to the Permanent Committee, and the Committee to the Joint Commission. As we also saw, working relations between the parties were exceptionally good during the 1990s, and the higher-up levels seldom had objections to what was presented from 'below'. As long as there was agreement at the expert level, the administrative level (and partly political level, to the extent that the heads of national delegations consulted with politicians at home) approved of going further. Finally, when the issue was presented to the Joint Commission after agreement had been reached first in the sub-group and then in the Permanent Committee – after several years of preparation, at that – the Commission saw no reason to put a halt to the initiative. It should be noted that the technical regulation was not accorded high priority in the Commission, despite the remark by the Norwegian delegation leader at the time that he viewed this as the major achievement from his stint as delegation leader (see Chapter 4). Until the harvest control rule was implemented in 2002, the heads of delegation spent most of their time during sessions in the Commission agreeing on the TAC and the exchange of quotas. Proposals from the Permanent Committee were dealt with rather quickly: in practice, adopted without much discussion. So, with the selection grids, agreement between Norwegian and Russian experts in the sub-group on selection technology, and in the Permanent Committee, was pivotal. If there was any scepticism among the higher levels of the Russian delegation to the Joint Commission, the established procedures raised the threshold for blocking formal agreement. Again, it had become standard operating procedure for the Commission to adopt measures that had been agreed in the Permanent Committee.

In sum, I maintain that the reason why Russia complied with its international obligation to conduct fisheries management according to the precautionary approach was *not* because this was in its declared interest, presumably not even its *perceived* interest (and hardly as a general cooperative gesture to other states). Quite the contrary, Russia followed suit *more or less unwillingly*, with Norway at the wheel. Transnational seafaring norms and good-neighbourly relations may have tuned the negotiators in to a pro-compromise wavelength, but I argue that institutional factors are best suited to explaining Russia's compliance. In the Barents Sea fisheries management, Russia gradually spun itself into an institutional web of continuously more elaborate decision-making procedures, with Norway taking the leading role after the end of the Cold War. Partly, the established formal and informal standard operating procedures led to decisions that the Russians were soon to criticize, although they stuck to them. Independently from this, there was in the Joint Commission a 'drive towards compromise' that might to some extent have overshadowed strictly defined national interests, or at least led the parties to interpret such interests as positively as they could, weighing them up against the possibility of reaching agreement. Compromise became the institutional hallmark of the Joint Commission. Finally, the role of individual leaders was important. Many difficult questions were solved by the two heads of delegation, more or less in private. At least in the Russian delegation, internal legitimacy was sometimes secured after the decision had been made. The role of post-agreement bargaining in all this will be further explored below.

WHY DO FISHERS COMPLY?

I claimed above that the specific 'request areas' in the Svalbard Zone, which fishers were requested to stay out of owing to too much intermingling of fry, could be interpreted as the result of bargaining between the Norwegian enforcement body and the (mostly) Russian fishing fleet. Coast Guard inspectors put considerable effort into convincing the fishers to stay out of these areas, but were also ready to adjust the exact shape of the 'request areas' (extension as well as allowable fishing depths) in line with information provided by the Russian side. Further, I noted that there was not much room for bargaining during inspection. The Norwegian inspectors meticulously followed their enforcement instructions and were generally perceived by the Russian fishers as 'hard to please'. On the other hand, it was practically impossible to cheat in the Norwegian zones of the Barents Sea, owing to the very strict enforcement regime there.

In Chapter 2, we saw that the literature on compliance in fisheries has emerged from stylized deterrence models which claim that people comply with the law when this is in their economic interest (Becker, 1968). In recent decades, however, this approach has been supplemented by the 'enriched' model of compliance, which basically claims that compliance is a far more complicated issue than assumed in the 'formal' deterrence-based models. Young (1979), Sutinen et al. (1990) and Tyler (2006) have brought in normative issues such as the individual's personal morality, peer pressure and the legitimacy of rules and procedures. Gezelius (2002, 2004, 2006, 2007) has refined the 'enriched' model of compliance in fisheries by focusing on the interdependence between norms and deterrence. For instance, he found that in one empirical setting formal enforcement was necessary for informal social control (Gezelius, 2004). He has called for greater attention to the Durkheimian compliance mechanism, which emphasizes the symbolic meaning of enforcement, as opposed to the Hobbesian and Habermasian mechanisms, with their emphasis on deterrence and rational communication (Gezelius, 2007).

We do not have detailed information about Russian fishers' compliance in the Barents Sea. The statistics of the Norwegian Coast Guard indicate that most fishers comply with most regulations most of the time: less than 20 per cent of inspections (of vessels of all nationalities) reveal violations of some sort; less than 5 per cent reveal serious violations (Hønneland, 1998, 2000a; see also Chapter 4). On the other hand, Norwegian authorities partly documented, partly estimated Russian overfishing of some 50 per cent of the national Russian cod quota in the Barents Sea both in the early 1990s and around the mid-2000s. If these estimates are correct, far more than 5 per cent of the north-west Russian fishing vessels must have been involved in overfishing. The main reason why this is not reflected in the statistics of the Norwegian Coast Guard is probably the fact that the vessels involved in overfishing would fish primarily in the Russian EEZ and the former Grey Zone, both beyond the mandate of the Norwegian Coast Guard. There is also some doubt as to the ability of Norwegian inspectors to reveal underreporting (which is an indication of overfishing), since physical inspection of a fishing vessel's holds is an extremely challenging affair (see Chapter 5). So the question of why Russian fishers comply must be somewhat diluted: why do *most* Russian fishers *appear to be in compliance* with Norwegian law *when they are inspected* by the Coast Guard?

In my interviews with Russian fishers, the message came through loud and clear: they simply do not dare to cheat, because of the strict enforcement regime in Norwegian waters. Above all, the meticulous inspection procedures of the Norwegian inspectors, and their competence and incorruptibility were emphasized. This points in the direction of a deterrence-based explanation, which I will not dismiss. However, there might be more to it. With the possible exception of a few fishers who focused primarily on the discriminatory nature of Norwegian inspections ('if the mesh is even a tiny bit smaller than what's allowed, they immediately define it as a violation'), the Norwegian Coast Guard seemed to enjoy a rather high degree of legitimacy among Russian fishers. Even those who criticized the Coast Guard inspectors for not listening to the Russian arguments characterized the Norwegian enforcement regime as 'effective'. Most of them would also conclude that 'this is the right thing to do'. Among the spontaneous replies to my questions about how the Russian fishers assessed the work of the Norwegian Coast Guard were the following: 'Good job!' and 'Top grade!' Several interviewees presented the Norwegian enforcement system as a case for emulation for the Russian enforcement bodies, which my interviewees, almost without exception, dismissed as incompetent and corrupt. Related to this, many interviewees hinted at normative factors such as the perceived need to preserve the fish stocks for future generations. The very positive descriptions given of the work of marine science filled in this picture. So did the hints at the existence of a 'seafaring community': 'We respect each other here at sea. We all have hard work.'

As discussed in Chapter 5, we cannot be sure whether the accounts offered by the Russian fishers in interviews are in fact 'genuine'. Were the descriptions of an effective Norwegian enforcement system and of utterly incompetent Russian enforcement bodies merely reproductions of old myths about Western order and Russian chaos? Were my interviewees just playing the 'saying all the right things' game when they 'innocently' spoke of the need to think of future generations? From my own observations and several hundred interviews in Russia (for other research projects), I would say that the answer lies somewhere in between. I think there are in most interviews glimpses of actual experience, and genuine perceptions. But I also think that many interviews to some extent reflect the 'standard operating procedures' – or socially established narratives – of the interview situation, or talk on these themes more widely. The best I can do with the present interview material is to conclude cautiously that Russian fishers seem to comply with Norwegian law partly because of the strict enforcement regime (deterrence) and partly because it is seen as 'the right thing to do' (norms). Somewhat similarly to Gezelius (2004) with the assumption that formal enforcement is necessary for (informal) social control, I believe that the formal enforcement regime in the Norwegian waters in the Barents Sea is a precondition for ('conservationist' or seafaring) norms to influence fishers' behaviour. Knowing that violations will most likely be discovered and punished presumably makes it easier to settle for a 'lawful bargain' oneself – to let one's own norms about right and wrong, or about proper fisher behaviour, come to the foreground of consciousness. In a different situation, with no effective control on the Norwegian side in the Barents Sea, I can imagine it would have been easier for many fishers to 'forget about the norms' and instead focus solely on their own short-term economic gains. Whether this explanation is deterrence- or norm-based is a matter of interpretation; I wish to emphasize the interrelationship between the two. At least in this specific empirical setting, it would have been difficult to rely on norms alone (unlike many small-scale fishing situations; see Chapters 1 and 2), but *legitimacy and communication serve to support the deterrence pillar of the enforcement system*. The potential gain from communication (or bargaining) is particularly evident in the Coast Guard's efforts to reduce fishing in areas with a lot of small fish. By contrast, it does not seem to be the case during the Norwegian inspections of Russian vessels, a point to which we return below.

POST-AGREEMENT BARGAINING REVISITED

I concluded above that Norway had considerable success in its attempts to influence Russian behaviour in the Joint Commission through post-agreement bargaining, largely through bargaining at lower bureaucratic (or technical or scientific) levels and in direct contact between the heads of delegation. I then took a slight detour in discussing *why* the Russian side so often adapted its behaviour according to Norwegian initiatives, which brought Russia closer to the precautionary standards of international fisheries law. *Russia unwillingly followed suit*, with Norway at the wheel. It was difficult to explain Russia's behaviour through realist models, although defensive realist and neoclassical realist approaches allowed for 'accidental' behaviour not necessarily in the declared interest of the state in question. Normative factors like traditional seafaring norms might have prepared the ground for compromise, but I argued that institutional factors could best explain the Russian compliance. In particular, Russia found itself in an institutional web of increasingly more elaborate decision-making procedures, geared largely towards *compromise*. Post-agreement bargaining was the practical tool employed by Norway to get new measures (such as selection grids) and procedures (such as the harvest control rule) implemented. In line with Chayes and Chayes (1995, p. 109), instances of apparent non-compliance were regarded as 'problems to be solved, rather than as wrongs to be punished' ('We've taken up the things we felt were wrong'). Post-agreement bargaining – between technical experts, by the heads of delegation in private, around dinner tables and in Russian *banyas* – was used, deliberately or by accident, *to activate norms and establish standard operating procedures that furthered precautionary fisheries management*. Post-agreement bargaining, then, emerges not as a source of compliance, but as a means to activate such sources: interest, norms or institutional features. Intentionally or not, Norway activated

Russian (or seafaring) norms and created institutional arrangements suited to precautionary fisheries management through post-agreement bargaining.[14] Theoretically, my study provides support for the normative and institutionalist perspectives on compliance and shows the potential of post-agreement bargaining to activate norms and develop standard operating procedures that further the objectives of the regime in question.

At the individual level, I concluded that Russian fishers mostly comply with Norwegian law when fishing in the Norwegian waters of the Barents Sea for two main reasons: they fear sanctions, and they see the Norwegian enforcement arrangement as just. Here we see reflected the symbolic role of enforcement, accentuated by Gezelius (2007). Enforcement has a role beyond deterrence: it acts to reassure citizens that law and order reign. As also argued above, *legitimacy and communication support the deterrence pillar* of the enforcement system. The apparent success of the Norwegian Coast Guard in persuading the Russian fishers to stay out of the 'request areas' of the Svalbard Zone shows the potential of argumentation – or bargaining: the inspectors were also willing to listen to the Russians' arguments. Here the inspectors appeared as *politicians*, in Kagan and Scholz's (1984) theoretical framework. In the inspection situation, however, the Russian fishers obviously perceived the Norwegian inspectors as *policemen*, geared only towards the disclosure and sanctioning of violations. We see hardly any trace of the inspectors as *consultants* aimed at solving practical problems, or as *politicians*, who, according to Kagan and Scholz (1984, p. 68), should have the competence 'to suspend enforcement, to compromise, to seek amendments to the regulations'. (Post-agreement) bargaining was a central element of the Coast Guard's working relations with the fishing fleet (directly to influence fishers' behaviour in specific situations, and indirectly to affect their norms and perceptions of an effective enforcement system) – *except during actual inspection*.

We see a conspicuous difference in how the potential of post-agreement bargaining is applied in the Barents Sea fisheries at state and at individual levels. At the state level, Norwegian scientists, technical experts and civil servants have indeed acted as 'politicians' and 'consultants', aiming at compromise and practical problem-solving ('We have consciously tried to show understanding for their problems, always'). However, according to the Russian fishers that I interviewed, such openness was not what characterized their contact with Norwegian enforcement bodies during inspection ('They act without wasting any words. And if they find anything, they won't listen to our explanations'). It could be argued that this comes as no surprise: civil servants act on behalf of politicians, while inspectors carry police authority. However, my observation of the interaction between the Norwegian Coast Guard and the Russian fishing fleet in defining areas suitable for fishing demonstrates the potential for 'political' or 'consultative' work also at this level. The sustainable practice of staying out of areas with too much fry would arguably not have materialized without the open and constructive dialogue between inspectors and fishers. While more flexibility on the part of the Coast Guard regarding minor violations is probably not compatible with current Norwegian law, a stronger focus on practical problem-solving during inspection might in fact have enhanced compliance.

More generally, international fisheries collaboration and national fisheries enforcement have a great untapped potential in traditional seafaring norms, the 'deeply rooted sense of belonging which can hardly be ascribed reason alone' (Gezelius, 2002, p. 313). Activating this potential through post-agreement bargaining may contribute significantly to making fishery agreements work.

NOTES

1. Russian enforcement statistics would have been necessary to provide even more sturdy conclusions about individual compliance, but these are not publicly available.
2. Short interview extracts are provided merely to illustrate a point. They are just repetitions of extracts from interviews already presented in Chapters 4 and 5, so references are not provided here.
3. This negotiation track was used not only in the Commission, but also in the Permanent Committee, albeit more seldom. A member of the Norwegian delegation to the Committee said of one particular Russian delegation leader: 'He was rather difficult. We often had to take him into the back room' (interview in Bergen, June 2011).
4. This may be a justification with hindsight, though. Norway did not have many alternatives to accepting the Russian demand for a TAC far above scientific recommendations in 1999.
5. As explained in Chapter 4, the Russian civilian enforcement body has appeared under different names during the past decade, reflecting reorganizations at the federal level: first Murmanrybvod (as in Soviet times) up until 2004, then the Federal Veterinary Service from 2004 to 2007 (when Russian fisheries management was the remit of the Ministry of Agriculture) and since 2007 BBTA, the regional branch of the Federal Fisheries Agency.
6. As mentioned, I do not go into detailed discussions about the *level* of compliance; without doubt, however, the Norwegian initiatives pulled the Russians in a direction that involved *increased* compliance with international fisheries law, such as precautionary setting of the TAC.
7. We saw in Chapter 4 that Norwegian negotiators believed that Russia's status as a great power set limits to what small-state Norway could achieve ('It was a matter of pride – the great power, and then the little one that came dragging in a lot of muck'). Whether realism provides a suitable explanation of this is more uncertain. By no means did Russia put any military or economic pressure on Norway when the latter produced documentation of Russian overfishing in the mid-2000s. (Admittedly, Russia imposed a ban on import of Norwegian salmon in early 2006, which by some was seen as a response to the Norwegian arrest of a Russian fishing vessel in the Svalbard Zone in the autumn of 2005; see Chapter 3.) It was more a matter of Russia being 'insulted' about

getting a 'scolding' from its little neighbour in the north-west. Perspectives on the role of identity in international relations may be more relevant (Neumann, 1996; Hønneland, 2010). There were limits to how far Russia could go in adapting to Norwegian demands before losing its perception of itself as the big brother in the relationship.

8. When the Federal Border Service took over responsibility for enforcement in the Russian EEZ in 1998 (see Chapter 3), this body was immediately given two seats in the Russian delegation to the Permanent Committee. However, the representatives from the Federal Border Service were shunned by the rest of the Russian delegation; I myself observed how they constantly approached the head of the Norwegian delegation with their requests, instead of going through the head of their own delegation. Their relations with the rest of the Russian fisheries complex gradually improved, but the institutional struggle continued. According to one of my interviewees, as we saw above, the Federal Border Service was never given access to the data about Russian landings in Norway that the Norwegian Directorate of Fisheries had been forwarding to the Russian civilian fisheries enforcement body since 1993.

9. At the infamous 1999 session of the Joint Commission (see Chapter 4), I was told by one member of the Russian delegation that the fishing industry had 'bought control of' the Russian delegation. The 'Russian interest' was thus defined by those who were willing to pay for it.

10. Generally positive characteristics were, however, often mentioned by the Russians, such as 'we handle the management of our joint fish resources successfully'.

11. There has traditionally been a strong 'sector identity' in Russia, including in fisheries (Hønneland, 2004). Up until recently, most people had the same educational background (mostly as fishery biologists) and alternated between different positions in the fisheries complex (science, regulation, enforcement or the fishing industry). Since the turn of the millennium, there has been a tendency to place people from the country's 'power structures' in the top positions of the fisheries management system, but most civil servants have still climbed the ladder of the fisheries complex, including service at sea.

12. The seafaring theme is also very noticeable at the Commission as a social and cultural arena, with delicious seafood dinners, boat trips and maritime excursions, as well as cultural programmes with a nautical twist.

13. Several prominent members of the national delegations to the delimitation negotiations were also members of the Joint Commission. Perhaps the 'rush for compromise' spilled over from the Commission to the delimitation negotiations (Hønneland, 2011).

14. The Norwegians undoubtedly had some intention of influencing Russian behaviour, but they might not have been overly aware of which mechanisms – or sources of compliance – they activated. The gradual introduction of selection grids is an example of a process that was probably unintentional on the Norwegian side: the Norwegians did not introduce selection grids in the shrimp fishery in order to make it easier for Russia to accept grids in the cod fishery some years later. Similarly, the stepwise adoption of new regulatory measures – from joint exploration in technical sub-groups via discussions in the Permanent Committee to formal adoption in the Joint Commission – was definitely applauded by the Norwegians, but hardly devised deliberately to make it easier to get their way vis-à-vis the Russians.

References

Aasjord, B. and G. Hønneland (2008), 'Hvem kan telle "den fisk under vann"? Kunnskapsstrid i russisk havforskning', *Nordisk Østforum*, **22**, 289–312.

Acheson, J.M. (1975), 'Fisheries management and social context: the case of the Maine lobster fishery', *Transactions of the American Fisheries Society*, **104**, 653–68.

Allison, G.T. and P.D. Zelikow (1999), *Essence of Decision: Explaining the Cuban Missile Crisis*, New York: Longman.

Andresen, S., E.L. Boasson and G. Hønneland (eds) (2012), *International Environmental Agreements: An Introduction*, London and New York: Routledge.

Axelrod, R. (2006), *The Evolution of Cooperation* (with a new foreword by Richard Dawkins), New York: Basic Books.

Baland, J.-M. and J.P. Platteau (1996), *Halting Degradation of Natural Resources: Is There a Role for Rural Communities?*, Oxford: Clarendon Press.

Becker, G. (1968), 'Crime and punishment: an economic approach', *Journal of Political Economy*, **72**, 169–217.

Bentham, J. (1789), *An Introduction to the Principles of Morals and Legislation*, London: T. Payne and Son.

Berenboym, B.I., V.A. Borovkov, V.I. Vinnichenko, E.N. Gavrilov, K.V. Drevetnyak, Yu.A. Kovalev, Yu.M. Lepesevich, E.A. Shamray and M.S. Shevelev (2007), 'Chto takoe sinopticheskiy monitoring treski v Barentsevom more?', *Rybnye resursy*, **4**, 24–9.

Berkes, F. (ed.) (1989), *Common-Property Resources: Ecology and Community-Based Sustainable Development*, London: Belhaven Press.

Bonger, W.A. (1916), *Criminality and Economic Conditions*, Boston, MA: Little, Brown and Co.

Borisov, V.M., S.I. Boychuk, G.P. Vanyushin, A.D. Gomonor, D.N. Klyutshkov, B.N. Kotenev, G.G. Krylov and B.M. Shatokhin (2006), *Sinopticheskiy monitoring zapasov treski v Barentsevom more v 2005 g. na osnove ispol'zovaniya sovremennykh issledovatel'skikh tekhnologiy izucheniya bioresursov*, Moscow: VNIRO Publishing.

Bose, S. and A. Crees-Morris (2009), 'Stakeholder's views on fisheries compliance: an Australian case study', *Marine Policy*, **33**, 248–53.

Bromley, D.W. (general ed.) (1992), *Making the Commons Work: Theory, Practice, and Policy*, San Francisco, CA: Institute for Contemporary Studies Press.

Brox, O. (1990), 'The common property theory: epistemological status and analytical utility', *Human Organization*, **49**, 227–35.

Burgstaller, M. (2005), *Theories of Compliance with International Law*, Leiden and Boston, MA: Martinus Nijhoff.

Chayes, A. and A.H. Chayes (1991), 'Compliance without enforcement: state behavior under regulatory treaties', *Negotiation Journal*, **7**, 311–30.

Chayes, A. and A.H. Chayes (1995), *The New Sovereignty: Compliance with International Regulatory Agreements*, Cambridge, MA and London: Harvard University Press.

Churchill, R. and G. Ulfstein (1992), *Marine Management in Disputed Areas: The Case of the Barents Sea*, London and New York: Routledge.

Cooter, R. and S. Marks, with R. Mnookin (1982), 'Bargaining in the shadow of the law: a testable model of strategic behavior', *Journal of Legal Studies*, **11**, 225–51.

Crawford, B.R., A. Siahainenia, C. Rotinsulu and A. Sukmara (2004), 'Compliance and enforcement of community-based coastal resource management regulations in North Sulawesi, Indonesia', *Coastal Management*, **32**, 39–50.

Dawes, R.M. (1975), 'Formal models of dilemmas in social decision making', in Martin Francis Kaplan and Steven Schwartz (eds), *Human Judgment and Decision Processes: Formal and Mathematical Approaches*, New York: Academic Press, pp. 88–107.

Eggert, H. and A. Ellegård (2003), 'Fishery control and regulation compliance: a case for co-management in Swedish commercial fisheries', *Marine Policy*, **27**, 525–33.

Franck, T.M. (1990), *The Power of Legitimacy among Nations*, New York and Oxford: Oxford University Press.

Garcia, S.M. (1994), 'The precautionary principle: its implications in capture fisheries management', *Ocean and Coastal Management*, **22**, 99–125.

Gardner, R., E. Ostrom and J. Walker (1990), 'The nature of common-pool resource problems', *Rationality and Society*, **2**, 335–58.

Gergen, K. (2001), 'Self-narration in social life', in Margaret Wetherell, Stephanie Taylor and Simeon J. Yates (eds), *Discourse Theory and Practice: A Reader*, Thousand Oaks, CA and London: SAGE, pp. 247–60.

Gezelius, S.S. (2002), 'Do norms count? State regulation and compliance in a Norwegian fishing community', *Acta Sociologica*, **45**, 305–14.

Gezelius, S.S. (2003), *Regulation and Compliance in the Atlantic Fisheries: State/Society Relations in the Management of Natural Resources*, Dordrecht and Boston, MA: Kluwer Academic.

Gezelius, S.S. (2004), 'Food, money, and morals: compliance among natural resource harvesters', *Human Ecology*, **32**, 615–34.

Gezelius, S.S. (2006), 'Monitoring fishing mortality: compliance in Norwegian offshore fisheries', *Marine Policy*, **30**, 462–9.

Gezelius, S.S. (2007), 'Three paths from law enforcement to compliance: cases from the fisheries', *Human Organization*, **66**, 414–25.

Glaser, B.G. and A.L. Strauss (1967), *The Discovery of Grounded Theory: Strategies for Qualitative Research*, Chicago, IL: Aldine.

Gordon, H.S. (1954), 'The economic theory of a common-property resource: the fishery', *Journal of Political Economy*, **62**, 124–42.

Gray, W.B. and J.T. Scholz (1993), 'Does regulatory enforcement work? A panel analysis of OSHA enforcement', *Law and Society Review*, **27**, 177–213.

Gubrium, J.F. and J.A. Holstein (2009), *Analyzing Narrative Reality*, Thousand Oaks, CA and London: SAGE.

Habermas, J. (1984), *The Theory of Communicative Action*, Boston, MA: Beacon Press.

Hardin, G. (1968), 'The tragedy of the commons', *Science*, **162**, 1243–8.

Hatcher, A. and D. Gordon (2005), 'Further investigations into the factors affecting compliance with U.K. fishing quotas', *Land Economics*, **81**, 71–86.

Hatcher, A., S. Jaffry, O. Thébaud and E. Bennett (2000), 'Normative and social influences affecting compliance with fishery regulations', *Land Economics*, **76**, 448–61.

Hauck, M. (2008), 'Rethinking small-scale fisheries compliance', *Marine Policy*, **32**, 635–42.

Hauck, M. and M. Kroese (2006), 'Fisheries compliance in South Africa: a decade of challenges and reform 1994–2004', *Marine Policy*, **30**, 74–83.

Henkin, L. (1968), *How Nations Behave: Law and Foreign Policy*, London: Pall Mall Press.

Henriksen, T. and G. Ulfstein (2011), 'Maritime delimitation in the Arctic: the Barents Sea Treaty', *Ocean Development and International Law*, **42**, 1–21.

Hersoug, B. (2005), *Closing the Commons: Norwegian Fisheries from Open Access to Private Property*, Delft: Eburon.

Hewison, G.J. (1996), 'The precautionary approach to fisheries management: an environmental perspective', *International Journal of Marine and Coastal Law*, **11**, 301–32.

Hoel, A.H. (2005), 'The performance of exclusive economic zones: the case of Norway', in Syma A. Ebbin, Alf Håkon Hoel and Are K. Sydnes (eds), *A Sea Change: The Exclusive Economic Zone and Governance Institutions for Living Marine Resources*, Dordrecht: Springer, pp. 33–48.

Hønneland, G. (1993), *Fiskeren og allmenningen; forvaltning og kontroll: Makt og kommunikasjon i kontrollen med fisket i Barentshavet*, Tromsø: University of Tromsø, Department of Social Science.

Hønneland, G. (1998), 'Compliance in the Fishery Protection Zone around Svalbard', *Ocean Development and International Law*, **29**, 339–60.

Hønneland, G. (1999a), 'A model of compliance in fisheries: theoretical foundations and practical application', *Ocean and Coastal Management*, **42**, 699–716.

Hønneland, G. (1999b), 'Co-operative action between fishermen and inspectors in the Svalbard Zone', *Polar Record*, **35**, 207–14.

Hønneland, G. (1999c), 'The stories fishermen tell (or: Themes from the Barents Sea fisheries)', *Human Ecology*, **27**, 621–6.

Hønneland, G. (1999d), 'The interaction of research programmes in social science studies of the commons', *Acta Sociologica*, **42**, 193–205.

Hønneland, G. (2000a), *Coercive and Discursive Compliance Mechanisms in the Management of Natural Resources: A Case Study from the Barents Sea Fisheries*, Dordrecht and Boston, MA: Kluwer Academic.

Hønneland, G. (2000b), 'Compliance in the Barents Sea fisheries: how fishermen account for conformity with rules', *Marine Policy*, **24**, 11–19.

Hønneland, G. (2000c), 'Enforcement co-operation between Norway and Russia in the Barents Sea fisheries', *Ocean Development and International Law*, **31**, 249–67.

Hønneland, G. (2003), *Russia and the West: Environmental Co-operation and Conflict*, London and New York: Routledge.

Hønneland, G. (2004), *Russian Fisheries Management: The Precautionary Approach in Theory and Practice*, Leiden and Boston, MA: Martinus Nijhoff.

Hønneland, G. (2005), 'Fisheries management in post-Soviet Russia: legislation, principles and structure', *Ocean Development and International Law*, **36**, 179–94.

Hønneland, G. (2006), *Kvotekamp og kyststatssolidaritet: Norsk–russisk fiskeriforvaltning gjennom 30 år*, Bergen: Fagbokforlaget.

Hønneland, G. (2010), *Borderland Russians: Identity, Narrative and International Relations*, Basingstoke and New York: Palgrave Macmillan.

Hønneland, G. (2011), 'Kompromiss als Routine: Russland, Norwegen und die Barentssee', *Osteuropa*, **61**, 257–69.

Jackson, P.T. (2010), *The Conduct of Inquiry in International Relations: Philosophy of Science and Its Implications for the Study of World Politics*, London and New York: Routledge.

Jacobson, H.K. and E.B. Weiss (1995), 'Strengthening compliance with international environmental accords: preliminary observations from a collaborative project', *Global Governance*, **1**, 119–48.

Jensen, Ø. (2011), 'Current legal developments, the Barents Sea: treaty between Norway and the Russian Federation concerning maritime delimitation and cooperation in the Barents Sea and the Arctic Ocean', *International Journal of Marine and Coastal Law*, **26**, 151–68.

Jentoft, S. (1985), 'Models of fishery development: the cooperative approach', *Marine Policy*, **9**, 322–31.

Jentoft, S. (1989), 'Fisheries co-management: delegating government responsibility to fishermen's organizations', *Marine Policy*, **13**, 137–54.

Jentoft, S. (2005), 'Fisheries co-management as empowerment', *Marine Policy*, **29**, 1–7.

Jentoft, S. and B.J. McCay (1995), 'User participation in fisheries management', *Marine Policy*, **19**, 227–46.

Jentoft, S., M. Bavinck, D.S. Johnson and K.T. Thomson (2009), 'Fisheries co-management and legal pluralism: how an analytical problem becomes an institutional one', *Human Organization*, **68**, 27–38.

Jönsson, C. and J. Tallberg (1998), 'Compliance and post-agreement bargaining', *European Journal of International Relations*, **4**, 371–408.

Jørgensen, A.K. (2009), 'Recent developments in the Russian fisheries sector', in Elena Wilson Rowe (ed.), *Russia and the North*, Ottawa: University of Ottawa Press, pp. 87–106.

Jørgensen, J.H. and G. Hønneland (2006), 'Implementing global nature protection agreements in Russia', *Journal of International Wildlife Law and Policy*, **9**, 1–21.

Kagan, R.A. and J.T. Scholz (1984), 'The "criminology of the corporation" and regulatory enforcement strategies', in Keith Hawkins and John Michael Thomas (eds), *Enforcing Regulations*, Boston and The Hague: Kluwer-Nijhoff Publishing, pp. 67–95.

King, D.M. and J.G. Sutinen (2010), 'Rational noncompliance and the liquidation of northeast groundfish resources', *Marine Policy*, **34**, 7–21.

King, D.M., R.D. Porter and E.W. Price (2009), 'Reassessing the value of U.S. Coast Guard at-sea fishery enforcement', *Ocean Development and International Law*, **40**, 350–72.

Koh, H.H. (1997), 'Why do nations obey international law?', *Yale Law Journal*, **106**, 2599–659.

Krebs, R.R. and P.T. Jackson (2007), 'Twisting tongues and twisting arms: the power of political rhetoric', *European Journal of International Relations*, **13**, 35–66.

Krementsov, N. (1997), *Stalinist Science*, Princeton, NJ: Princeton University Press.

Kuperan, K. and J.G. Sutinen (1998), 'Blue water crime: deterrence, legitimacy, and compliance in fisheries', *Law and Society Review*, **32**, 309–37.

McCay, B.J. and J.M. Acheson (eds) (1987), *The Question of the Commons: The Culture and Ecology of Communal Resources*, Tucson: University of Arizona Press.

McCay, B.J. and S. Jentoft (1996), 'From the bottom up: participatory issues in fisheries management', *Society and Natural Resources*, **9**, 237–50.

Mearsheimer, J.J. (2001), *The Tragedy of Great Power Politics*, New York: Norton.

Mitchell, R.B. (1994a), 'Regime design matters: intentional oil pollution and treaty compliance', *International Organization*, **48**, 425–58.

Mitchell, R.B. (1994b), *Intentional Oil Pollution at Sea: Environmental Policy and Treaty Compliance*, Cambridge, MA and London: MIT Press.

Mnookin, R.H. and L. Kornhauser (1979), 'Bargaining in the shadow of the law: the case of divorce', *Yale Law Journal*, **88**, 950–97.

Morgenthau, H. (1948), *Politics among Nations: The Struggle for Power and Peace*, New York: Alfred A. Knopf.

Neumann, I.B. (1996), *Russia and the Idea of Europe: A Study in Identity and International Relations*, London and New York: Routledge.

Neumann, I.B. (2008), 'Discourse analysis', in Audie Klotz and Deepa Prakash (eds), *Qualitative Methods in International Relations: A Pluralist Guide*, Basingstoke and New York: Palgrave Macmillan, pp. 61–77.

Nielsen, J.R. (2003), 'An analytical framework for studying: compliance and legitimacy in fisheries management', *Marine Policy*, **27**, 425–32.

Nielsen, J.R. and C. Mathiesen (2003), 'Important factors influencing rule compliance in fisheries: lessons from Denmark', *Marine Policy*, **27**, 409–16.

Nøstbakken, L. (2008), 'Fisheries law enforcement: a survey of the economic literature', *Marine Policy*, **32**, 293–300.

Olson, M. (1965), *The Logic of Collective Action: Public Goods and the Theory of Groups*, Cambridge, MA: Harvard University Press.

Ostrom, E. (1990), *Governing the Commons: The Evolution of Institutions for Collective Action*, Cambridge and New York: Cambridge University Press.

Ostrom, E., R. Gardner and J. Walker (1994), *Rules, Games, and Common-Pool Resources*, Ann Arbor: University of Michigan Press.

Pedersen, T. (2008), 'The constrained politics of the Svalbard offshore area', *Marine Policy*, **32**, 913–19.

Pedersen, T. (2009a), 'Norway's rule on Svalbard: tightening the grip on the Arctic islands', *Polar Record*, **45**, 147–52.

Pedersen, T. (2009b), 'Denmark's policies toward the Svalbard area', *Ocean Development and International Law*, **40**, 319–32.

Pedersen, T. (2011), 'International law and politics in U.S. policymaking: the United States and the Svalbard dispute', *Ocean Development and International Law*, **42**, 120–35.

Pinkerton, E. (ed.) (1989), *Co-operative Management of Local Fisheries: New Directions for Improved Management and Community Development*, Vancouver: University of British Columbia Press.

Pyle, D.J. (1983), *The Economics of Crime and Law Enforcement*, London: Macmillan.

Randall, J.K. (2004), 'Improving compliance in U.S. federal fisheries: an enforcement agency perspective', *Ocean Development and International Law*, **35**, 287–317.

Ries, N. (1997), *Russian Talk: Culture and Conversation during Perestroika*, Ithaca, NY and London: Cornell University Press.

Ringmar, E. (1996), *Identity, Interest and Action: A Cultural Explanation of Sweden's Intervention in the Thirty Years War*, Cambridge: Cambridge University Press.

Røttingen, I., H. Gjøsæter and B.H. Sunnset (2007), 'Norsk–russisk forskersamarbeid 50 år', *Havforskningsnytt*, **16**, Bergen: Institute of Marine Research.

Rubin, H.J. and I.S. Rubin (1995/2005), *Qualitative Interviewing: The Art of Hearing Data*, 1st and 2nd editions, Thousand Oaks, CA and London: SAGE.

Sen, S. and J.R. Nielsen (1996), 'Fisheries co-management: a comparative analysis', *Marine Policy*, **20**, 405–18.

Serebryakov, V. and P. Solemdal (2002), 'Cooperation in marine research between Russia and Norway at the dawn of the 20th century', *ICES Marine Science Symposia*, **215**, 73–86.

Smith, A. (1759), *The Theory of Moral Sentiments*, London: A. Millar.

Smith, A. (1776), *An Inquiry into the Nature and Causes of the Wealth of Nations*, London: Strahan & Cadell.

Somers, M. (1994), 'The narrative constitution of identity: a relational and network approach', *Theory and Society*, **23**, 605–49.

Southall, T., P. Medley, G. Hønneland, P. MacIntyre and M. Gill (2010), *MSC Sustainable Fisheries Certification: The Barents Sea Cod and Haddock Fisheries*, Inverness: Food Certification International.

Spector, B.I. and I.W. Zartman (eds) (2003), *Getting It Done: Post-Agreement*

Negotiation and International Regimes, Washington, DC: United States Institute of Peace Press.

Sproule-Jones, M. (1982), 'Public choice theory and natural resources: methodological explication and critique', *American Political Science Review*, **76**, 790–804.

Stokke, O.S. (2009), 'Trade measures and the combat of IUU fishing: institutional interplay and effective governance in the Northeast Atlantic', *Marine Policy*, **33**, 339–49.

Stokke, O.S. (2010a), *A Disaggregate Approach to International Regime Effectiveness: The Case of Barents Sea Fisheries*, Oslo: Unipub.

Stokke, O.S. (2010b), 'Barents Sea fisheries: the IUU struggle', *Arctic Review on Law and Politics*, **1**, 207–24.

Stokke, O.S. (forthcoming), *Disaggregating International Regime Effectiveness: Theory, Method, Governance*, Cambridge, MA and London: MIT Press.

Sutinen, J.G. and P. Andersen (1985), 'The economics of fisheries law enforcement', *Land Economics*, **61**, 387–97.

Sutinen, J.G. and K. Kuperan (1999), 'A socioeconomic theory of regulatory compliance in fisheries', *International Journal of Social Economics*, **6**, 174–93.

Sutinen, J.G., A. Rieser and J.R. Gauvin (1990), 'Measuring and explaining noncompliance in federally managed fisheries', *Ocean Development and International Law*, **21**, 335–72.

Tyler, T.R. (2006), *Why People Obey the Law* (with a new afterword by the author), Princeton, NJ: Princeton University Press.

Underdal, A. (2000), 'Conceptual framework: modelling supply of and demand for environmental regulation', in Arild Underdal and Kenneth Hanf (eds), *International Environmental Agreements and Domestic Politics: The Case of Acid Rain*, Aldershot and Burlington, VT: Ashgate, pp. 49–86.

Vylegzhanin, A.N. and V.K. Zilanov (2007), *Spitsbergen: Legal Regime of Adjacent Marine Areas*, Utrecht: Eleven International Publishing.

Walt, S. (1998), 'International relations: one world, many theories', *Foreign Policy*, **110**, 29–45.

Weber, M., G. Roth and C. Wittich (1978), *Economy and Society: An Outline of Interpretive Sociology*, Berkeley: University of California Press.

Weiss, E.B. and H.K. Jacobson (eds) (1998), *Engaging Countries: Strengthening Compliance with International Environmental Accords*, Cambridge, MA and London: MIT Press.

Wendt, A. (1992), 'Anarchy is what states make of it: the social construction of power politics', *International Organization*, **46**, 391–425.

Wilson, D.C., J.R. Nielsen and P. Degnbol (eds) (2003), *The Fisheries Co-management Experience: Accomplishments, Challenges and Prospects*, Dordrecht and Boston, MA: Kluwer Academic.

Young, O.R. (1979), *Compliance and Public Authority: A Theory with International Applications*, Baltimore, MD and London: Johns Hopkins University Press.

Index